Research and
Development Abroad
by U.S. Multinationals

Robert Ronstadt

The Praeger Special Studies program—utilizing the most modern and efficient book production techniques and a selective worldwide distribution network—makes available to the academic, government, and business communities significant, timely research in U.S. and international economic, social, and political development.

Research and Development Abroad by U.S. Multinationals

PRAEGER SPECIAL STUDIES IN INTERNATIONAL ECONOMICS AND DEVELOPMENT

Praeger Publishers   New York   London

Library of Congress Cataloging in Publication Data

Ronstadt, Robert, 1942-
    Research and development abroad by U.S.
multinationals.

    (Praeger special studies in international
economics and development)
    Bibliography: p.
    1. Research, Industrial.   2. Technology
transfer.   3. Corporations, American.   I. Title.
HD30.4.R66   607'.2'73   77-10672
ISBN 0-03-022661-9

PRAEGER SPECIAL STUDIES
200 Park Avenue, New York, N.Y., 10017, U.S.A.

Published in the United States of America in 1977
by Praeger Publishers,
A Division of Holt, Rinehart and Winston, CBS, Inc.

789   038   987654321

© 1977 by Praeger Publishers

Printed in the United States of America

For Becky

ACKNOWLEDGMENTS

I owe a particular debt of gratitude to Professor Robert B. Stobaugh, of the Harvard Business School. His ability to pose the right questions at the right time made this project extremely rewarding for me.

A special note of appreciation is also due to Professor Richard Rosenbloom, Sarnoff Professor of the Management of Science and Technology at the Harvard Business School. His suggestions and good research advice helped to make my field research work a much richer experience for me than it would have been without his council.

Because the program for the Management of Science and Technology is jointly run by the Harvard Business School and the Massachusetts Institute of Technology, I was fortunate to have Professor Thomas Allen of MIT advise me. His participation deserves special thanks. The encouragement he offered when I encountered difficulties helped me to overcome them.

My acknowledgments would certainly be incomplete without mentioning the thoughtful council of Dr. Steven Whitelaw, whose numerous suggestions aided me in organizing several chapters of this study.

I also wish to acknowledge the assistance provided by Professor Joseph Bower of the Harvard Business School, who helped considerably with one of the field research projects, and whose practical advice in conducting certain interviews proved very helpful to me.

Financial support for this study was provided generously by the Division of Research of the Harvard Business School and Babsow College. The Multinational Enterprise Project of New York University also lightened the financial burden of research.

No doubt exists in my mind that the study owes its existence to the exemplary cooperation of over 50 executives from more than a dozen enterprises. My only hope is that this work reflects the fine efforts of these individuals.

Finally, the study would not have been possible without the strong support provided by my wife, Rebecca Ronstadt, to whom this work is dedicated.

CONTENTS

LIST OF TABLES AND FIGURES

In June 1976, *Business Week* introduced its annual survey of U.S. R&D spending with the fascinating, yet all too accurate, observation about the R&D process:

> Corporate executives, economists, engineers, and scientists all broadly agree that the nation's growth and security depend directly on the ill-defined range of activities called research and development. Despite this consensus, the picture of how R&D works, where it is done, and who does what is murky.[1]

The objective of this study is to make the water less murky for a relatively small but growing segment of R&D activity: that portion of R&D performed abroad by U.S.-based multinational enterprises. At least some tentative answers are provided regarding "how R&D works" abroad, "where it is done, and who does what. . . ." Equally important, some reasons are also presented that explain "why" R&D investments are created abroad and why they tend to evolve in certain directions.

This kind of knowledge about the process of making R&D investments abroad is important for two reasons: First, some evidence exists from this study that, when combined with other studies being conducted by the author, suggests that the domestic process of creation and evolution of R&D investments is very similar to the R&D process experienced abroad. The implication is that a U.S. multinational based in New York can make an R&D investment in Texas for much the same reasons as for an R&D investment made in France. Moreover, the pattern of evolution of these R&D investments is likely to be similar. Consequently, knowledge about the "foreign" R&D process may be useful for some managers when analyzing the R&D commitment of a U.S. multinational.

Second, the explanations and descriptions of the creation and evolution of R&D investments abroad are potentially important for some organizations right now regarding their operations abroad. While R&D investments abroad are not large in the aggregate, some multinationals already have significant and expanding commitments of R&D resources outside the United States.

For instance, by 1974, IBM was spending in the neighborhood of $200 million abroad for R&D (roughly 30 percent of its total R&D budget), considerably more than most technology-intensive firms were spending in the United States.

Even among some "small" R&D spenders, foreign R&D activity was significant. For example, Otis Elevator was allocating nearly half of its total R&D dollars abroad in 1974. By 1973, CPC International was spending approximately 38 percent of its total R&D outlays abroad. Exxon Corporation (including chemicals) was spending $25 million for R&D abroad, approximately one fourth its total budget.

Moreover, these investments in R&D abroad appeared to have some significant consequences for each of these multinational systems. For example, the contributions of IBM's foreign labs in the development of the 360 and 370 computer lines are documented and well known. Less known, but equally noteworthy, is that most of IBM's R&D work in voice communications has been done abroad in its French and Swiss labs. In the case of elevators, most of the recent technology transferred by Otis Elevator to lower income nations has come not from the United States but from Otis' European subsidiary, based on R&D work performed in Europe for small elevators for small to medium-sized buildings.

The emerging direction of technology transfers was similar for CPC International. Also, some new technology developed by CPC Europe was being transferred to the United States. As one CPC manager noted, "There is little doubt that on the food side of our business, the best R&D work has been done by our affiliate in Europe."* Finally, a new role for R&D abroad was emerging for some companies as foreign markets grew in size relative to the U.S. market. This new role was perhaps most clearly stated by an R&D manager at Exxon Chemical:

> At one time almost all new chemical plants were started first in the United States, but this is changing now because Europe as a whole is almost equivalent of the U.S. market in chemicals. Since Exxon Chemical now has R&D and all other functions existing in Europe, you may see more initial commercialization in Europe for chemical products which previously may have been commercialized in the United States.

These observations about the magnitude and impact of R&D abroad give rise to a number of questions with potentially important repercussions for businessmen, economists, and public policy makers. For example, what is the impact of R&D investment abroad on the efficient allocation of R&D resources in a multinational system? Are certain kinds of foreign R&D activities more successful than others? Do some types of foreign R&D activity present more opportunities or raise more problems when they are established or as they evolve than

---

*All nonreferenced quotes are from private interviews the author had with managers from the seven multinationals participating in this study.

others? What kind of impact do different kinds of foreign R&D activities have on international transfers of technology by U.S.-based multinational enterprises?

SCOPE AND METHOD

The foreign R&D activities of seven multinational organizations are analyzed in this study. These seven organizations include Exxon Corporation's energy businesses, Exxon Chemical Company, International Business Machines (IBM), Chemicals and Plastics Group of Union Carbide Corporation, CPC International, Otis Elevator Company, and Corning Glass Works. Each multinational system was analyzed as a total entity, except Exxon and Union Carbide. Data were collected separately for the Exxon Chemical Company and for Exxon's energy-related businesses because the Exxon Chemical Company was organizationally separate from Exxon's energy businesses (including R&D operations); and the motivations for investing abroad in chemicals versus the natural resource-based businesses were different. In the case of Union Carbide, the decision to study only the chemicals and plastics group was based on the complexity of the company, and the fact that the group's interactions with the other two major business groups within the firm were minimal.

The findings are based on interviews conducted by the author during 1974 with approximately 60 executives in these seven multinational organizations. The interviewees were selected for their knowledge of foreign R&D activities, but were not limited to R&D managers. The author checked data gathered at corporate headquarters and U.S.-based sites by visiting selected foreign R&D units. These foreign sites were different from each other in terms of the reasons for their formation and the number of years each had been in existence.

In total, the seven multinationals in this study created or acquired 55 R&D units. The following table lists these R&D units for each multinational:

Unit Number	Company
	Union Carbide Corporation (UCC)
1	UCC Netherlands (Antwerp) Unit
2	UCC Switzerland (Versiox) Unit
3	UCC British (Hythe) Unit (chemicals)
4	UCC Canada (Montreal) Unit (plastics)
5	UCC Canada (chemicals) Unit
6	UCC Indian (process) Unit
7	UCC Indian (New Delhi) Unit
8	UCC Australian Unit
9	UCC Canadian (Toronto) Unit (food casings)
10	UCC British (Tysly) Unit (BXL)
11	UCC British (Manningtree) Unit (BXL)

Unit Number	Company
12	UCC British (Scotland) Unit (BXL)
13	UCC Belgium (ERA) Unit

Corning Glass Works (COR)

14	COR British (Sunderland) Unit
15	COR French (Le Versinet) Unit
16	COR French (Soveril) Unit
17	COR French (ERL) Unit
18	COR British (Halstead) Unit

Otis Elevator Company (OTIS)

19	OTIS German Unit
20	OTIS British Unit
21	OTIS French Unit
22	OTIS Italian Unit
23	OTIS French Unit (Paris HQ)
24	OTIS Austrian Unit
25	OTIS Spanish (San Sebastian) Unit
26	OTIS Spanish (Madrid) Unit
27	OTIS French (Saxby) Unit

CPC International (CPC)

28	CPC Belgian Unit
29	CPC British Unit
30	CPC French Unit
31	CPC German (Krefeld) Unit
32	CPC German (Heilbronn) Unit
33	CPC Swiss (Knorr) Unit
34	CPC Italian Unit
35	CPC Japanese Unit

Exxon Corporation Energy Businesses (EXXP)

36	EXXP British Unit
37	EXXP French Unit
38	EXXP Canadian Unit
39	EXXP German Unit
40	EXXP Italian Unit

Exxon Chemical Company (EXXC)

41	EXXC British Unit
42	EXXC French Unit
43	EXXC Canadian Unit

Unit Number	Company
44	EXXC German Unit
45	EXXC Belgian Unit
46	EXXC Canadian (LaSalle) Unit

International Business Machines (IBM)

47	IBM British Unit
48	IBM French Unit
49	IBM German Unit
50	IBM Switzerland Unit
51	IBM Netherlands Unit
52	IBM Austrian Unit
53	IBM Canadian Unit
54	IBM Swedish Unit
55	IBM Japanese Unit

Data were aggregated for these 55 R&D units for two points in time: the year when they were created and 1974 or the year the R&D unit was disbanded. Each R&D unit was classified according to its primary R&D purpose when it was created or acquired and as it existed in 1974. Those characteristics associated with the R&D purpose of foreign R&D units were singled out. Data about these relationships were used to identify hypotheses that appeared to explain why these R&D units were created abroad and why they evolved in particular directions over time.

A number of variables were used to describe the characteristics of transfer technology units, indigenous technology units, global product units, and corporate technology units. Usually, these variables were rather straightforward, for example, the national location of the unit, its date of creation, its age in 1974, the number of R&D professionals employed when started in 1974, their national origin, where the lab was specifically located (at a manufacturing site, a national headquarters, a regional headquarters), and so on. However, three new variables were used in the course of this study in order to arrive at a better understanding of R&D activities in general and the foreign R&D operations of multinational enterprises specifically. These new variables involved the identification of R&D expenditures by their strategic purpose; the identification of R&D units by their administrative positions vis-a-vis other functional units; and the identification of R&D units by the geographic scope (national, regional, or global) of their R&D responsibility.

The first variable involved modifying definitions of R&D activity developed originally by the Industrial Research Institute (IRI).[2] As a result of initial fieldwork, the IRI definitions were rewritten and a new category was added. This category developed from discussions with R&D directors that led to the belief that if the IRI definition of "exploratory research" were subdivided, the nature of strategic choice would be more accurately defined, because when an exploratory research project is approved, R&D directors usually know whether the potential "product, process or service" will fall within existing business lines or lead to new high-risk business lines.

Consequently, four strategic categories of R&D expenditures are used to analyze the R&D operations of each multinational organization in this study.

1. *R&D that supports existing business*. This is R&D that is conducted to support the company's existing business—to maintain or improve its profitability or to improve its social acceptance. R&D is conducted to retain or to increase market share by introducing new products, by decreasing the cost of manufacture or preventing excess increases in manufacturing cost, by extending existing products into new applications, by enhancing safety, by reducing pollution, or by improving product or market acceptance in other ways.

2. *R&D that develops new high-risk business*. R&D to develop new high-risk business projects is conducted with the intention of developing a product or process in which the sponsoring company has no direct manufacturing, marketing, management experience, and technology. Included are projects that involve diversification into a new venture for the company or into a totally new way of accomplishing an important existing function. Such R&D is frequently high risk in nature. It may follow the successful accomplishment of exploratory research or may be a new program related to technology acquired in other ways. It may include any or all of the categories of technical work associated with research and development.

3. *Exploratory research that supports existing business*. Exploratory research that supports existing business is performed to advance knowledge of phenomena that are of general interest to the company or to find major new business projects that are within the company's existing business experience. Usually long range in nature, it may include literature searches, laboratory scouting experiments, and preliminary economic evaluation. A new product, process, or service is in view, but the work by definition remains exploratory until a specific product or process objective is established.

4. *Exploratory research that develops new high-risk business*. Exploratory research that develops new high-risk business is performed to advance knowledge of phenomena that are of general interest to the company or to find major new business projects that are outside the organization's direct experience in manufacturing marketing, management, and technology. Usually long range in nature, it may include literature searches, laboratory scouting experiments, preliminary application and engineering studies, and preliminary economic evaluation. A new product, process, or service is in view, but the work by definition remains exploratory until a specific product or process objective is established.

The second variable used to describe R&D operations is based on the administrative position of R&D units vis-a-vis other functional units in the company. Such analysis is needed because the administrative bonds, or barriers, between R&D and other functions play an important role in determining the type of activities performed by an R&D unit. In this study, an R&D unit's administrative position is identified as either subordinate or independent: subordinate if the R&D unit's director reports to the manager of a functional area other than R&D and independent if the unit's director reports to another R&D manager or to a general manager. If two or more reporting lines exist, the unit's organizational position may be subordinate in one case and independent in another. In this situation, the unit's organizational classification, subordinate or independent, depends on whether the majority of the unit's R&D professionals perform projects for the manager of a functional area outside R&D or for an R&D or a general manager.

The third variable used to analyze foreign R&D activities involved the classification of R&D units by their principal geographic market areas of responsibility. For instance, several R&D units in a multinational enterprise could be similar in some respects, yet perform the same R&D function for different geographic markets. In preliminary study, it was found that R&D projects were performed either for national (single nation), for regional (several nations), or for global (without specific geographic boundaries) markets. Consequently, an R&D unit's level of responsibility was determined by the level of responsibility held by the unit's director—whether he or she reported to a manager of a local (national) affiliate, a regional headquarters, a corporate headquarters, or two or more of the above.

The creation and evolution of foreign R&D investments can be explained in terms of the type of R&D activities performed abroad and/or other factors not related directly to the performance of R&D. For instance, included in the latter group are the following possible motivations for the establishment of R&D units outside the United States: to utilize "blocked" funds in a foreign country, to take advantage of "cheap" R&D talent, to obtain government incentives, to monitor other R&D groups, to curb the local "brain drain," and to satisfy foreign government policies or pressures.

These and other non-R&D-related factors were noted whenever it was found that they had been instrumental in the establishment of a foreign R&D unit. But their usefulness as analytic tools was limited since an infinite number of such factors could exist. Also, their existence alone did not explain why certain kinds of R&D activities were performed initially and over time.

The literature, as well as the author's exploratory fieldwork, suggested three R&D-related explanations for the creation of foreign R&D units.[3] to service and/or to adapt product/process technology transferred from the U.S. parent company to its foreign affiliates;[4] to generate new or improved products and/or processes expressly for foreign application;[5] and to stimulate technology flows from the foreign R&D unit to the U.S. parent for application in the United States and/or other countries (outside the country in which the R&D unit was located).[6]

Each of these explanations was supported by the observations of multinational managers involved in R&D activities. The principal reasons cited by these sources provided a starting point for further investigation into the foreign R&D phenomenon. But these reasons did not adequately explain why some multinational organizations created R&D units abroad and others did not. Nor did they explain why some organizations became involved in R&D abroad at particular times in certain countries and business areas but not at other times and places in the same businesses. In short, although the three reasons above described the purposes for which U.S.-based multinational organizations create R&D units abroad, they did not explain why these purposes had developed.

Research revealed the existence of four kinds of R&D investments abroad when defined by the purpose of their R&D activity. These were R&D units that primarily performed R&D to help transfer U.S. technology to a foreign market, called "transfer technology work"; performed R&D expressly for the foreign market, called "indigenous technology work"; performed R&D to develop new and improved products for simultaneous application in the United States and abroad, called "global product work"; and performed R&D expressly for the U.S. parent, called "corporate technology work." Within these four categories, an R&D purpose was defined as "primary" when the majority of R&D expenditures was committed to that purpose. For instance, an R&D unit was called a transfer technology unit when 51 percent (or a simple majority) of its R&D expenditures were for transfer technology work. However, the same R&D unit could become an indigenous technology unit if its budget for similar work became at least 51 percent (or a majority of the total expenditures).

At this point, one factor not mentioned in the literature must be introduced that had a bearing on the development of this system of classification. This factor was the acquisition of foreign R&D operations by U.S.-based multinational organizations. All previous explanations for the existence of foreign R&D units were based on the assumption that U.S.-based multinational organizations had some motive for initiating R&D activity abroad. In early fieldwork, it was found that some U.S.-based multinational organizations had also acquired foreign R&D operations unintentionally or incidentally. These units were part of a foreign company acquired by the U.S.-based parent, but in every instance the R&D capability of the acquired firm was not a reason for the acquisition by the parent company.

R&D units acquired incidentally were clearly different from those established (or intentionally acquired) by the parent company. The primary R&D purpose(s) of these incidentally acquired foreign R&D units might be unrelated to the transfer of the parent company's technology, the development of new or improved products, or generating technology flows from the R&D unit to the U.S. parent. For instance, an R&D unit that was acquired incidentally might simply be supporting a foreign business. The business might be similar to or different from the parent's but, in either case, not dependent on U.S. technology transfers.

Consequently, in the present study a distinction has been made between acquired R&D units and units created by the parent company. Also included as a primary R&D purpose is any R&D performed expressly for a foreign market (and not solely for the development of new or improved products and processes). It was convenient to refer to this kind of R&D activity as "indigenous technology" work.

NOTES

1. *Business Week*, "Where Private Industry Puts Its Research Money," June 28, 1976, p. 62.

2. See Alfred E. Brown, "New Definitions for Industrial R&D," *Research Management*, September 1972, pp. 55-57. As Brown and others have shown, the traditional definitions of R&D (for example, fundamental or basic, applied and development) do not reveal why the activities are performed, that is, their strategic implications.

3. Three main sources exist that describe several possible motivations for foreign R&D activities at the company level. See David B. Hertz, "R&D as a Partner in World Enterprise," in *McKinsey Anthology*, 1971, pp. 317-25; The Conference Board, *R&D in the Multinational Company: A Survey*, Reports for International Management, no. 8 (New York: Conference Board, 1970); and U.S., Congress, Senate, Committee on Finance, *Implications of Multinational Firms for World Trade and Investment and for U.S. Trade and Labor*, Chap. 6, pp. 550-604, February 1973, which includes some aggregate analysis of R&D spending at home and abroad by U.S. manufacturing multinationals for 1966.

At the country level of analysis, some excellent studies have been conducted: for Canada, Arthur J. Cordell, *The Multinational Firm, Foreign Direct Investment and Canadian Science Policy*, Special Study no. 22 (Ottawa: Science Council of Canada, December 1971); and A. E. Sefarian, *The Performance of Foreign-Owned Firms in Canada* (Montreal: Private Planning Association, 1969); for Great Britain, J. H. Dunning, *American Investment in British Manufacturing* (London: Allen & Unwin, 1958); for Germany, *U.S. Subsidiaries in the German Federal Republic* (New York: Commerce Clearing House, 1969).

In addition to these works as background, a number of exploratory interviews were conducted with a number of managers of five multinational organizations with corporate headquarters in the Boston area. The majority of these managers had personal experience with R&D operations abroad in their own companies as well as in other organizations where they had been employed.

4. See E. G. Hough, "Communication of Technical Information Between Overseas Markets and Head Office Laboratories," *Research Management*, March 1, 1973, pp. 1-5. See also Cordell, op. cit., p. 46, who refers to these types of R&D units as "support laboratories."

5. See Raymond Vernon, "The Location of Economic Activity," Working Paper, Harvard University, 1973, pp. 8-9, for a fuller discussion. Citations of various studies that indicate the existence of a market or demand bias in the innovation process are provided.

6. See E. G. Woodroofe, "Technology and Business Opportunity for International Business," in *Technological Change and Management*, ed. David W. Ewing (Boston: Harvard University Press, 1970). See also U.S., Congress, op. cit., p. 583; and Christopher Tugendhat, "IBM's International Integration," in *The Multinationals* (New York: Random House, 1972), where some of these same points are reiterated in a study of IBM's R&D operations in the United States and abroad.

THE EXPERIENCES
OF U.S. MULTINATIONALS
WITH R&D ABROAD

INTRODUCTION
TO PART I

Each of the next seven chapters summarizes the particular experiences of seven U.S. multinationals with R&D abroad. Each chapter presents information about the organization's overall R&D commitment, the kinds of R&D investments established abroad, how these same R&D investments evolved over time, and an in-depth look at some selected foreign R&D investments.

The particular order of presenting each organization's experiences with R&D abroad is arbitrary but does reflect the intensity of its R&D commitment abroad relative to its total R&D investment:

Estimated R&D Abroad in 1974 for
Seven U.S. Multinationals
(percent of total R&D)

Corning	9
Union Carbide (chemicals and plastics only)	12
Exxon Chemical	23
Exxon (energy only)	25
IBM	31
CPC International	39
Otis Elevator	45

Source: Company records and estimates made by company managers and author.

1

By 1974 the Corning Glass Works could view a history of considerable change as a technology company. The corporation had evolved from a small craft-based domestic firm into a science-based multinational enterprise with over 1,000 people in its R&D organization. For years R&D activities were centralized both administratively and geographically in Corning, N.Y. However, the basic R&D structure evolved into a mixed corporate and divisional line R&D structure by 1960 with the advent of domestic product divisions. R&D activities began a process of administrative and spatial decentralization that is continuing today with the advent of serious R&D operations abroad. However, the geographic decentralization of R&D has occurred primarily through the acquisition of domestic and foreign firms with ongoing R&D activities rather than through the creation of small service R&D units associated with investments in new manufacturing facilities in the United States or abroad. Data for 75 manufacturing facilities showed that 69 plants had no R&D units associated spatially with their operations on a permanent basis. The exceptions were all gained from acquisitions with preexisting R&D operations.

From 1955 to 1973, the number of R&D professionals more than doubled from 180 (all in the United States) to 425, with roughly 6 percent located abroad. R&D expenditures climbed sevenfold over the same period from $5 to $35 million, with 9 percent expended abroad (see Table 1.1).

TABLE 1.1

Corning Glass Works: Geographic Distribution of R&D Employment and Expenditures for Selected Years

Year	World Total	United States	Abroad[a]	Percent of Total Abroad
R&D Expenditures (in millions of dollars)				
1955	5.1	5.1	0	0
1960	9.4	9.4	0	0
1965	20.0	20.0	0	0
1970	26.1	24.0	2.1	7.8
1973	35.2	31.9	3.3	9.3
R&D Employment of Professionals[b]				
1955	180	180	0	0
1960	220	220	0	0
1965	300	300	0	0
1970	345	325	20	5.8
1973	425	400	25	5.9

[a] All R&D expenditures and employment were located in Europe.
[b] R&D professionals were defined by Corning as degreed scientists and engineers.
Source: Company records.

SUMMARIZING CORNING'S EXPERIENCE WITH
FOREIGN R&D ACTIVITIES

Corning's experience with R&D abroad involved the establishment of five foreign R&D units. Three of these five R&D units are transfer technology units; the other two are indigenous technology units.

The two foreign R&D units created by Corning that performed technology transfer work were a technical assistance unit created in 1967 by a newly established manufacturing subsidiary, Electrosil, Ltd., to modify electronic products and reduce costs and located in Sunderland, U.K.; a technical assistance unit created in 1971 by a newly established manufacturing subsidiary, SOVCOR, to modify electronic products and reduce costs and located in Le Vesinet, France. The strategic composition of R&D expenditures at both units when they were created was 100 percent in support of existing businesses. Each unit employed two R&D professionals who were nationals of their respective countries. The director of each unit reported directly to the general manager of each subsidiary.

The third foreign R&D unit was acquired incidentally by Corning. It was a technical service unit located in Avon, France, that was obtained in 1969 when Corning assumed majority ownership of a French specialty glass firm, Soveril (hereafter called the Soveril R&D unit).

The Soveril R&D unit was providing primarily technical assistance for some businesses based on technology obtained from Corning from licensed technology. In the past, the unit had worked on projects to generate new and improved specialty products and processes for the local market and had developed some specialized skills. However, its main function had been technical assistance work for the past several years before it was acquired by Corning. The Soveril R&D unit was performing 100 percent R&D in support of existing business activities when Corning acquired it. The unit was composed of ten R&D professionals, all of whom were French nationals. The director of the unit reported directly to the president of Soveril before and after Corning's acquisition.

Two foreign R&D units performed primarily indigenous technology work to generate new or improved products primarily for the European markets. One unit, created by Corning, an R&D unit called European Research Laboratories located in Avon, France that was created in 1974 by Corning's European regional headquarters (with strong support from corporate headquarters) to develop a self-sufficient R&D capability for Corning's growing European business. The unit would perform some technology transfer activities, but its principal function would be new product development evolving eventually into an R&D capability on a par with Corning's corporate R&D unit in the United States.

Within a short time after its creation, the unit began performing some exploratory research leading into existing business areas (20 percent) and R&D to develop new high-risk businesses (30 percent). The remaining 50 percent was

R&D in support of existing business activities. The unit employed 18 R&D professionals, all French nationals except the director, who was a U.S. citizen. The director reported directly to the president of Corning Europe. He also reported informally to corporate R&D executives in the United States.

A second R&D unit that performed indigenous technology work and that was acquired incidentally by Corning was located in Halstead, U.K. It was obtained in 1972 when Corning purchased 100 percent ownership of an English medical instruments company, Evans Electroselenium, Ltd. The British R&D unit was responsible for developing new and improved products for the medical instruments market. These products were supplementary to Corning's own line of medical instrument products. The unit performed 100 percent R&D in support of existing business operations. It was composed of one R&D professional who reported directly to the subsidiary's general manager.

By mid-1974, Corning's R&D operations abroad involved four R&D units. The French R&D unit acquired from Soveril had been disbanded in a legal and administrative sense. However, the unit's personnel were retained along with their laboratory facility in Avon as they became part of the newly created European R&D unit.

The two transfer technology units created by Corning subsidiaries in the electronics business experienced very little growth in the number of R&D professionals (each added one R&D professional). Also, neither unit experienced any change in its original foreign R&D function to help apply technology for electronic products supplied by the domestic parent. R&D expenditures remained 100 percent in support of existing business, and each director still reported solely to his subsidiary's general manager.

The Soveril R&D unit did not experience any significant growth in personnel until Corning decided to detach the unit legally and administratively from Soveril and make the unit part of its new R&D center for Europe. The unit was separated legally and administratively from Soveril because Corning's senior executives did not believe other European foreign subsidiaries would utilize the new unit's resources if it remained part of the French subsidiary. Since the European unit was created when the author was conducting field research, enough time had not elapsed for it to realize any significant changes in size or function.

The British indigenous technology unit acquired from Evans grew from one to three R&D professionals. Its foreign R&D function remained unchanged, that is, to develop new medical products for the local market. The strategic composition of their R&D expenditures did change slightly to include some R&D to develop new high-risk business (10 percent) of total outlays. The balance (90 percent) remained in support of existing business activities. The unit's director continued to report to the subsidiary's general manager.

CORNING'S DECISION TO CREATE A EUROPEAN
INDIGENOUS TECHNOLOGY UNIT

Corning's R&D activities are currently experiencing fundamental change. This change involves an expansion of R&D activities abroad. The expansion, however, is more pervasive than simply doing more of the same thing. If Corning's goals are realized, the result will be a basic change in the strategic composition of R&D resources abroad with attendant changes in R&D-sponsored technology flows across national borders.

The consolidation of six European subsidiaries into wholly owned operations and the establishment of a regional European office led to the decision to create a European R&D capability equivalent to U.S. R&D capability over the next dozen years. The motivating rationale was the recognition that 50 percent of Corning's business would be outside the United States by 1980.

The only existing R&D facility capable of handling this expansion was Soveril's R&D lab at Avon. Consequently, Corning became interested in expanding the activities of this lab to serve its future European needs. However, some important questions still remained regarding the national location of the lab, the composition of its personnel, and its management and organization. Some of the newly consolidated subsidiaries had been Soveril's rivals for many decades. Whether or not they would utilize the expanded services and capabilities of Soveril's lab was uncertain. And even if they did, no guarantee existed that the lab's management and personnel would not continue to treat Soveril's needs on a first-priority basis.

In order to serve not only Soveril's interests but the interests of the other European subsidiaries and eventually corporate-level interests, Corning decided to detach legally and administratively Soveril's R&D lab from Soveril and make it part of the European regional headquarters and an extension of Corning Research/United States.

A key R&D director with Corning's Technical Staff Division was chosen to head what would be a European lab at Avon. His long-term mandate was to perform R&D in Europe for Corning Glass Works worldwide. The initial goal, however, would be R&D work for the European subsidiaries. Organizationally, the director's job was defined to build a "research" organization (as opposed to one concerned primarily with product modification work), which over an eight- to ten-year period would be technically on a par with Corning's R&D in the United States. The director would report to the president of Corning/Europe but would also have "very, very strong dotted line and close working arrangements with the top two R&D executives in the United States. The new director of European R&D was also expected to participate actively in European marketing decision making."

The location of the principal site for the new European lab would remain the same. The French government and Soveril had made various concessions and commitments to allow the construction of the lab in the historic Avon/Fontaine-bleau area. Corning was determined to honor these agreements even though the R&D expansion decision was made in the mid-1960s before Corning gained control of Soveril. A modern new facility had been built about one and a half hours southeast of Paris, and 30 minutes from Soveril's main manufacturing operations. The French government permitted construction in an otherwise off-limits historic area because the city of Fontainebleau was experiencing a serious unemployment problem at the time. For its part, Soveril agreed to a timetable to lift the future lab's employment to capacity level. This timetable was altered with Corning's acquisition of Soveril. The lab employed about 80 people in 1974 (all French nationals) and had existing capacity for 250 persons. However, sufficient land was available at the site to expand capacity to 500 people if needed in the future.

Numerous changes were implemented after Corning acquired control of the Avon lab in 1969. The lab was reorganized around product lines rather than scientific specialties. Divisional financial support for technical requests was lowered as Corning provided a larger share of corporate funds in order to provide Soveril's divisions with a greater incentive to utilize the lab's facilities. Individual projects had been reduced, consolidated, and enlarged. Overall, Corning's U.S.-based R&D managers were pleased with the lab's progress.

CHAPTER

2

UNION CARBIDE CORPORATION:
ITS CHEMICALS
AND PLASTICS BUSINESSES

The chemicals and plastics (C&P) businesses of Union Carbide accounted for worldwide sales of $1.6 billion in 1973 (42 percent of Union Carbide's total sales volume). Official R&D expenditures for the chemicals and plastics group totaled $35.4 million, or 2.2 percent of sales (see Table 2.1). Approximately 99 percent of these 1973 R&D expenditures were allocated to R&D units located in the United States and only 1 percent abroad (about $200,000). According to these figures, expenditures for R&D and their geographic distribution had not changed appreciably since 1965. However, these figures were stated conservatively for accounting purposes by Union Carbide insofar as they excluded R&D performed by majority-held foreign affiliates owned less than 100 percent; technical assistance work performed by R&D professionals at both domestic and foreign R&D units; and a significant amount of R&D performed by a corporate-created R&D unit abroad that no longer existed in 1973. If these expenditures were included, the budgets of Union Carbide's foreign R&D units were estimated by Union Carbide managers to exceed $9 million in 1973 and represented approximately 12 percent of C&Ps total (adjusted) R&D outlays.

While only a small percentage of Union Carbide's R&D resources were allocated abroad by 1973, the corporation had first-hand experience with a variety of foreign R&D commitments. Within C&P, these R&D commitments included a corporate-sponsored research lab in Belgium, a process-service unit in Antwerp, several small units for customer technical service, and three new R&D labs in the United Kingdom obtained by acquisition.

9

TABLE 2.1

Union Carbide: Geographic Distribution of R&D Expenditures for Chemicals and Plastics for Selected Years

Year	World Total	United States	Abroad	Europe
R&D Expenditures (in millions of dollars)				
1960	31.9[a]	no records available		0.1
1965	35.0	34.8	0.1	0.1
1968	36.8	36.6	0.1	0.1
1970[b]	34.5	34.1	0.3	0.1
1972	31.1	30.9	0.1	0.1
1973	35.4	35.1	0.2	0.1
By Percentage of World Total				
1960	100	no records available		0.3
1965	100	99.4	0.3	0.3
1968	100	99.4	0.3	0.3
1970	100	98.8	0.9	0.3
1972	100	99.4	0.3	0.3
1973	100	99.1	0.6	0.3

[a]In 1960: In addition to charges developed within R&D, certain "development" charges from the manufacturing function were designated as R&D. There are no records in R&D that show the amount charged in 1960, but $1 to $1.5 million was an approximation.
[b]The year 1968 is the first after the formation of chemicals and plastics division where consolidated records and expenditure analyses were made. Prior to 1968, all data are reconstructed.
Source: Company records.

SUMMARIZING UNION CARBIDE'S EXPERIENCE
WITH FOREIGN R&D ACTIVITIES

Union Carbide's experience with R&D abroad in its chemicals and plastics businesses involved the establishment of 13 foreign R&D units. Twelve foreign R&D units performed exclusively transfer technology work when they were created or acquired by Union Carbide. One R&D unit was created as a corporate technology unit. Seven transfer technology units were created by Union Carbide and five were acquired abroad.

The seven R&D units created abroad by Union Carbide included a process technical assistance unit created by the Belgian affiliate in the late 1950s to service a polyethylene production facility and located in Antwerp at the production site; a product technical assistance unit created by regional headquarters in 1961 to service regional customer needs in plastics and located in Versoix, Switzerland, in order to be near regional headquarters and in a "neutral" country so that joint venture affiliates located countries belonging to the European Free Trade Association (EFTA) and to the European Economic Community (EEC) would utilize the lab's services; a chemicals product and process technical assistance unit created by the U.K. affiliate in the early 1960s to service the process needs of a chemicals plant plus customer product needs, and located in the United Kingdom at the production site (called the Hythe Lab); a technical assistance unit created by the Canadian affiliate in 1963 to service local product needs in plastics and located in Montreal at the production site; a process technical assistance unit created by the Canadian affiliate in the early 1960s to service a chemicals production facility and located at the production site; a process technical assistance unit created by the local Indian affiliate in 1973 to service the needs of an agricultural chemicals production facility and located at the production site; and a product technical assistance unit created by the local Indian affiliate in 1974 to service the agricultural chemical product needs of the national affiliate and located in New Delhi.

When these R&D units were created, the strategic composition of R&D expenditures was 100 percent in support of existing business activities for each foreign R&D unit. Each foreign R&D unit was composed of less than five R&D professionals who were host nationals, except the Swiss lab, which had eight R&D professionals at creation and a mixed European employment composition. Each unit's R&D manager reported to a functional manager outside R&D at the national level, except the Swiss unit whose director reported to a functional manager outside R&D but who had regional business responsibility.

Five foreign R&D units were acquired incidentally by Union Carbide. These included a technical assistance unit obtained in 1954 when an Australian company that produced plastics was acquired; a technical assistance unit obtained in 1963 and located in Toronto when a Canadian company that produced food casings was acquired; a plant technical assistance unit obtained in 1973 and located in Tysly, U.K., when Bakelite Xylonite Limited (BXL), a producer of plastic products, was acquired; a plant technical assistance unit obtained in 1973 and located in Manning Tree, U.K., when BXL was acquired; and a plant technical assistance unit obtained in 1973 and located in Scotland when BXL was acquired.

All five R&D units were classified as transfer technology units because they were performing manufacturing and customer technical assistance work that was based on technology supplied by Union Carbide to the respective acquisition companies *before* these companies were acquired. The strategic composition of R&D expenditures was classified as 100 percent in support of existing business operations when each unit was acquired by Union Carbide. The first two R&D units had only a few R&D professionals when Union Carbide acquired them. The three BXL units, however, had, respectively, 16, 10, and 20 R&D professionals when they joined the Union Carbide system. Their size was attributed to the fact that the units had existed for many years because Union Carbide began supplying BXL with technology in polyethylene over two decades ago. All R&D professionals were nationals of their respective countries. The director of each unit reported to functional managers outside of R&D at the local company level.

One corporate technology unit was created in 1956 by corporate headquarters in order to generate technology in chemicals expressly for the U.S. parent. The unit, called European Research Associates, was located in Brussels, Belgium, in order to utilize outstanding European scientific talent who could also monitor the European scientific community for new product/process possibilities (hereafter called the Belgian unit). The scientific foreign location was chosen because Brussels was a relatively centralized location for an operation designed to attract top scientists from various European countries yet whom Union Carbide wanted to remain in Europe.

The Belgian unit performed 100 percent exploratory research in support of new high-risk business activities. Within a short time after creation, the unit had 37 R&D professionals who were from several European countries, except the director who was a U.S. citizen. The unit's director reported directly to top corporate officials in New York and was organized as a separate legal entity unattached to any other Union Carbide subsidiary abroad.

Two of Union Carbide's 13 foreign R&D units were disbanded by mid-1974: the Hythe, U.K. unit, which was created as a transfer technology unit, and the Belgian unit, which was created as a corporate technology unit. All 12 R&D units created or acquired as transfer technology units continued to perform technology transfer activities as their primary R&D purpose until 1974 or when disbanded. However, the Swiss R&D unit created by regional managers had started performing indigenous technology activities to develop new and improved products for the European region.

The Antwerp unit grew very little with only a half dozen R&D professionals in 1973. The unit continued its involvement in technical assistance for plant process work.

The Swiss R&D unit grew from 20 people and 8 R&D professionals to about 60 people and 23 R&D professionals in 1974 because of market growth realized by wholly owned Union Carbide affiliates in plastics, process chemicals, and coating intermediates; and the transfer of the Hythe lab's R&D personnel in chemicals to the Swiss unit. Transfer technology activities grew roughly from 8 to 13 R&D professionals, with the remaining personnel shifted into product development work for the European businesses. Approximately 40 percent of the Swiss R&D unit (10 R&D professionals) had evolved into activities that were oriented toward generating new and improved products for the European market.

The Hythe, U.K., unit expanded its activities so that it was sufficiently large by 1963 to merit a separate facility at the affiliate's chemicals plant site. By 1968, the unit became increasingly concerned with chemicals technical assistance projects for other European countries. However, the feeling emerged at Union Carbide's European headquarters in Geneva that these European projects could be handled better and more economically from one of the existing R&D units located on the Continent. Consequently, the Hythe unit was disbanded in 1970 and most of its personnel (eight or nine R&D professionals) were transferred to the Swiss R&D unit.

The Canadian plastics R&D unit employed approximately 15 R&D professionals by 1974. Its function remained the same: tailoring products to meet customer needs. The Canadian chemicals R&D unit employed 45 R&D professionals by 1974 who were responsible for plant technical assistance at UCC Canada's numerous manufacturing facilities.

The two R&D units created by the Indian affiliate were recently established (1973) and had not yet changed size or purpose by 1974.

The strategic composition of R&D expenditures for the surviving units had not changed by 1974 but remained 100 percent R&D in support of existing businesses. The Hythe unit also performed 100 percent R&D in support of existing businesses until it was disbanded. R&D directors continued to report to functional managers outside R&D at the subsidiary level, although informal reporting arrangements existed with senior R&D managers in the United States. The Swiss R&D unit remained the only foreign R&D unit with regional R&D responsibility.

Among the R&D units acquired by Union Carbide that were performing primarily transfer technology work was an Australian unit that grew from a few R&D professionals in the mid-1950s to 25 R&D professionals in 1974. Its primary R&D function during this period remained the same: to perform technical support for chemical products and processes transferred to Australia by Union Carbide. The unit had R&D expenditures amounting to $1 million in 1974, which were 100 percent in support of existing businesses. The unit's director continued to report to a plant manager.

Another was the Toronto R&D unit, which employed about eight R&D professionals in 1974. The unit's function continued to involve only technical support work for the food casings business. The strategic composition of R&D was 100 percent in support of existing businesses. The unit's director reported to a functional manager outside R&D at the local level.

No substantial change had occurred by 1974 at the three BXL units since they were acquired by Union Carbide in 1973. The units experienced no change in the number of R&D personnel. The strategic composition of R&D expenditures ($3 million) for the three units remained 100 percent in support of existing businesses. The directors of each unit continued to report to functional managers outside R&D at the subsidiary level.

The evolution of the Belgian R&D unit experienced no significant change in the size of its corporate technology activities until it was disbanded in 1969. The unit never became involved in transfer technology or indigenous technology activities. Its strategic composition remained 100 percent exploratory research to develop new high-risk business until operations were terminated, although part of this was oriented toward existing business activities during the unit's last few years. The unit's director continued to report to senior corporate managers until the unit was disbanded.

AN IN-DEPTH LOOK AT THREE UNION CARBIDE R&D INVESTMENTS ABROAD

In spring of 1974, R&D managers in the chemicals and plastics operations division were involved in planning R&D strategy for future chemicals and plastics activities. A key issue was the future role of R&D operations outside the United States. Their attention was focused particularly on Europe and India.

In Europe, an existing chemicals and plastics R&D unit was expanding its facilities in Switzerland. Also, three additional R&D units were being added from Bakelite Xylonite Ltd. (BXL), now a wholly owned subsidiary. In India, the decision involved the feasibility of creating an R&D capability in agricultural chemicals and, if feasible, where the lab should be located and what its responsibilities should be. The extent, nature, and location of the European and Indian R&D operations were being studied carefully by chemicals and plastics managers. Just a few years ago, these same managers had suffered an agonizing experience with an earlier R&D investment in Europe. The lab, located in Brussels and called European Research Associates, encountered a variety of problems and was finally disbanded in 1969.

The Brussels Lab: A Corporate Technology Unit

Union Carbide's research lab was dedicated in Brussels in 1956. The lab was established as a distinct legal entity unattached to any other Union Carbide subsidiary abroad and the lab's director reported directly to Union Carbide's U.S. corporate management.

The original motivation for locating the lab in Europe was to tap European scientific talent and "monitor" the European scientific community for innovative possibilities leading to new products and processes for Union Carbide. The source of this motivation was corporate headquarters and top domestic R&D managers. The operating divisions opposed the establishment of the Brussels lab as well as the other corporate domestic research labs being established at the same time. However, corporate leaders felt the corporation had to grow with new products or face eventual decline. The operating divisions were concerned primarily with their existing businesses and seemed to be drifting increasingly toward the production of commodities, especially in chemicals, with low and/or shrinking profit margins.

The Brussels lab reached its operating level of 180 people (37 PhDs) soon after dedication. The lab had a much higher ratio of technicians and support personnel than U.S. research labs because they were cheaper and more available and the lab's director felt existing ratios were too small and cut into productivity of professionals (ratio was three technicians to every PhD, or 111 technicians). Initially, research labor costs were about half what they were in the United States. In the 1960s, the lab's budget ran about $1.8 million.

Despite these favorable cost factors, the lab's productivity in terms of promoting Union Carbide's interests eventually came into question. After a dozen years of research activity, the Brussels lab could point to no direct contribution to the corporation. According to chemicals and plastics managers and the Brussels lab's own management, the lab's poor productivity record was due to:

1. A lack of defined goals for research. Corporate headquarters had not specified particular goals or boundaries of research activity.

2. The actual probability of discovering a business area for Carbide was very small under these circumstances. Many discoveries were made but they did not fall within the market horizon of company officials and/or were not sufficiently promising to persuade management to go into totally unrelated businesses. A new discovery usually meant that the corporation had no production, marketing, or general management skills to apply to the new project. Corporate headquarters inevitably concluded they did not know enough about this new business.

3. Even if a project successfully generated discoveries within the existing scope of Union Carbide's business framework, no one was looking to see if a market really existed for potential new products. With hindsight, managers felt the probability of finding a random fit between a discovery and a market need was infinitesimal. They believed the procedure should have been just the opposite, that is, identify market needs first and then direct research at a broad, lasting market need even though the research was of a long-term, exploratory nature.

At first Union Carbide's U.S. R&D managers thought they could rectify these problems by providing goals and project direction that fell within the company's existing business activity. But the lab could not be turned around:

> The real underlying problem was that we had hired 37 PhDs, most of whom were Europeans who had never worked for Union Carbide. What we thought we had hired was 37 innovators. PhDs are not necessarily innovators, and even if some were innovators, they were not familiar with Carbide and so they were not going to automatically innovate things we could use. Academically and scientifically, the lab was a great success. They published all kinds of articles. Their goals were being satisfied. Why should they change?

Divesting the lab was not an easy process due to Belgian law, nor was it a pleasant undertaking. It took time, money, and guts to admit the investment was a mistake. Union Carbide's senior R&D managers in chemicals and plastics had no desire to relive the experience.

The Swiss Lab: A Transfer Technology Unit

Union Carbide's European headquarters was established in Geneva during the late 1950s. The decision to create a technical service center in Versoix just outside Geneva stemmed from a desire to have it located near regional offices. However, the decision was also determined by the need for a central, neutral country because the lab was designed initially to service the needs of three joint venture subsidiaries in different countries, two of them in EFTA and one in EEC. These subsidiaries were Unifos in Sweden, BXL in Great Britain, and a Belgian firm.

Apparently, the creation of a neutral site did not stimulate the use of the technical center's services, since Unifos and BXL have never used Versoix's resources but have developed their own in-house R&D capability.

Versoix began operations in 1961 as a plastics laboratory with administrative ties to the polyethylene production facility located in Antwerp. At the time, no more than 15 to 20 people and 10 R&D professionals were assigned to the lab.

In 1959 Carbide created a new wholly owned subsidiary in the United Kingdom, Union Carbide Ltd., to produce chemicals. A small R&D unit doing technical service for customers evolved, and in 1963 these activities were sufficiently large to merit a separate facility at the plant site (known as the Hythe Laboratory). In 1968, Union Carbide Europe assigned a technology director to run this lab, which, by 1969, was becoming increasingly concerned with European chemical projects. In 1969 the decision was made to consolidate the Hythe lab with the Versoix lab. At first, the Hythe lab was going to be consolidated with the Brussels corporate/technology unit, European Research Associates. However, when corporate officers decided to disband ERA, the Hythe operation was moved to Versoix.

The Versoix lab does not support all of Union Carbide's operations in Europe. Union Carbide Europe is divided into nine business centers. The business centers supported by the Versoix lab are the Coating Intermediates Business Center, the Plastics Business Center, and the Process Chemicals Business Center. Each business center is headed by a product director with profit responsibility. In 1973, the three business centers had total sales close to $400 million. The Versoix lab provided technical support for roughly $100 million. However, 30 percent of the $100 million were imported sales from the United States with the remainder, approximately $70 million, for products manufactured in Europe. One product (polyethylene) in the Plastics Business Center accounted for a high share of technical support in terms of European manufactured sales.

The Versoix lab's budget in 1974 was divided into three principal kinds of activity: customer technical service (50 percent), product application and development (30 percent), and plant technical assistance or process work performed mainly in Antwerp (20 percent). The entire budget was spent for R&D in support of existing businesses (approximately $1.3 million in 1974). The lab's director noted:

> No real research is performed in Europe. I would call it mostly technical service, some product application and product development work. But to get and hold good people, the lab must move deeper into its technology area. It's not enough to do technical service. Versoix will attract good R&D personnel by getting into more product application work.

Approximately 23 product application and development projects were then underway at Versoix. The principal beneficiaries of this work would be Carbide's national companies in Western Europe and a few Eastern European customers. Over the last four years (1970-73 inclusive), the principal users of the lab's successful projects were located in Switzerland, the United Kingdom, Germany, Italy, Benelux countries, and Eastern Europe. One project, a new brake fluid formulation, was transferred and used successfully in the United States.

Approximately 60 people work at the lab. The national composition is extremely varied with 15 different nationalities from Western Europe, Eastern Europe, and North America. The lab's director is the only U.S. citizen. The lab's occupational composition is shown in Table 2.2.

The Versoix lab's group (that is, project) leaders have marketing counterparts in Geneva. They must mutually work out and agree on what R&D projects

TABLE 2.2

Union Carbide: Occupational Composition of the
Versoix Lab Staff, 1974

Composition	Number	Notes
Chemists/engineers	23	3 PhDs, mostly Western Europeans; business language is English
Chemists'/engineers' assistants	19	Business language is French or German
Secretaries/receptionist and librarian	5	Mostly Swiss
Stores/purchasing and maintenance	6	
Apprentices (assistants)	2	Part-time school/work program—Swiss
Total	55	

Source: Company interview.

will be performed and what each project's budget will be for the following year in terms of manpower needs. The parent company has little control over the lab's budgetary process. U.S./R&D managers do not review specific projects and the lab is formally independent from the parent. The lab's director reports primarily to the product director of the European Coatings Intermediates Business Center. although the lab also performs work for two other European business centers. However, informal communication ties exist with U.S. R&D managers through associations developed by the lab's director when he managed R&D programs in the United States.

One of the director's major goals was to formalize these associations by making the Versoix lab part of the domestic R&D project planning and budgeting process. This would allow Versoix to channel needed longer range European projects to the United States and yet maintain control over these projects even though R&D was being performed in the United States. As the lab's director noted:

> All this is done informally now. It depends on who you know. R&D managers in the United States, however, are generally budgeted out so that European work was either bootlegged or some special funds were budgeted through U.S. operations via International contribu-

tions. Europe's voice was heard sometimes but even then the timing was often off because the projects were not controlled from Europe.

At the end of March 1974, the lab director and the director of Coatings Intermediates Products attended an R&D directors' meeting in New York. A tentative agreement was reached on budgeting procedures for projects that Versoix might wish to fund through U.S. chemicals and plastics R&D units. Two main types of projects were identified: those involving products exported from the United States and not manufactured in Europe and those involving products that were manufactured abroad. In the former, U.S. profit center managers would have control and the power to approve or disapprove the project. Where products were manufactured in Europe, Versoix would control the project. The U.S. profit center manager would be informed of the project but would not have approval or disapproval powers.

R&D planning in Union Carbide Europe was a difficult process because a considerable amount of the region's sales still come from U.S. imports. Only polyethylene and urethanes were manufactured in Europe by Union Carbide. Much more R&D was performed in Europe by Versoix for these two product lines. Polyethylene and urethanes also had higher sales than the imported products. But even if total sales from imports from the United States were higher, more R&D in Europe would not be performed according to Versoix's R&D managers because these import sales were not constant. One R&D group leader remarked that:

> The import faucet goes on and off and it is hard if not impossible to get R&D budget commitments from European product directors in these lines when they cannot be sure of next year's general sales level.

One reason for this uncertainty resided in the "balancing" nature of these imports. When demand by the U.S. market accounted for most or all of existing capacity, the balance of exports for Europe was small or nonexistent. Price alone did not determine who received the product. For instance, Union Carbide Europe may be willing to pay $.40 a pound for chemical X, but if Union Carbide had heavy volume commitments with RCA in the United States for $.19, then RCA obtained the product. Good economic reasons may exist for such a policy. The point is that R&D was often a multiyear activity and sales volumes based on trade often did not provide the stability needed to fund local R&D operations, even when R&D resources and capability already existed in the region in the form of an ongoing R&D unit.

The problem, then, was that Union Carbide Europe's high trade component and relatively small manufacturing base made R&D expansion difficult, especially into product application and development work. This limited both marketing and R&D capability. For instance, Versoix faced tremendous R&D competition, primarily from Bayer in polyurethane foams. Versoix had no more than a few R&D professionals working in the polyurethane foam area, while Bayer had over 100 people. While Versoix has an inventory of U.S. technology available to it, the lab's director still felt it could not compete well. The result was that Versoix stayed out of certain areas within the polyurethane area.

The New Delhi Lab: The Decision to Create a Transfer Technology Unit in the Far East in Agricultural Chemicals

Senior R&D managers in chemicals and plastics were faced with another R&D investment decision: should they permit the establishment of an R&D laboratory in New Delhi, India, to support the growing agricultural chemicals business. Domestic R&D managers believed the decision would be precedent setting if an R&D lab were created in India. Other chemicals and plastics business areas in the Far East and other regions might begin lobbying for R&D capability. How far was chemicals and plastics willing to decentralize the performance of R&D operations? The decision, in short, was forcing the determination of policy regarding the worldwide location of R&D resources.

The Indian/Far East argument for an R&D lab was persuasive. It rested on the contention that the creation and continuing support of a viable agricultural chemicals business required the performance of several major R&D activities in India. These included biological and field testing for new and existing products, metabolism and residue analysis, product development and formulation, process chemistry and process development, and a synthesis program for new products.

Proponents for the Indian lab admitted the United States could provide the Indian agricultural chemical business with all future R&D support. However, they argued that such support would be expensive and time consuming, especially for field testing, metabolism and residue analysis, and product development and formulation for the Indian market. One R&D manager noted:

> I don't believe that letters between the U.S. Charleston lab and India can adequately describe the product and formulation needs which depend so much on local conditions, be it cultural or economic, and for this reason, I believe that formulation and product development efforts for the Indian markets can be more efficiently supported in India.

According to the laboratory's supporters, the inclusion of a process chemistry and process development capability would be essential to enable local servicing of agricultural chemicals plants in order to make economic short-term changes in production operations. Furthermore, a process R&D capability in India would improve the international transfer of process technology by (1) speeding the transfer, thereby helping to avoid critical construction delays that could result in insufficient capacity by the time the plant came on stream (as actually happened once to a petrochemical facility in Brazil); and (2) providing India with the capability of developing an optimum process technology for India given its resources, domestic labor costs, and new technological developments occurring long after the original U.S. technology was developed. In the latter case, smaller Indian market sizes meant that India received older, obsolete U.S. process technology. As Figure 2.1 demonstrates, the time gap could be substantial. A local Indian process R&D unit could determine the feasibility of bringing the technology up to date and/or more in line with Indian economic circumstances, especially if it was located at the Indian agricultural chemical production site.

FIGURE 2.1

Union Carbide: International Process Technology Transfer
Between the United States and India—An Illustration of Time Gap

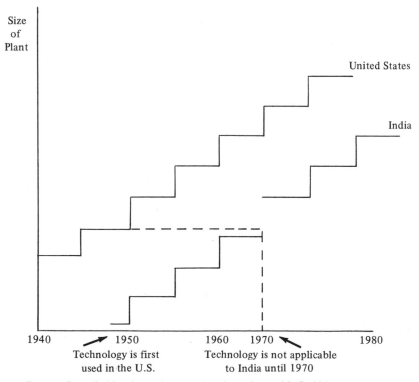

Source: Compiled by the author based on interviews with Carbide managers.

Eventually, a chemical synthesis group for new products would be created once a self-supporting R&D/agricultural chemical business was realized. By then, the New Delhi lab would be capable of supporting not only the R&D needs of the Indian subcontinent but also of other low-income nations in the Far East region and possibly in the African and Latin American regions.

In the meantime, the potential Indian market was substantial and R&D competition in India was minimal. According to one manager, the situation was similar in some ways to an R&D decision made much earlier in Union Carbide's history:

> I believe it is important to view Carbide's early R&D lab output dur-
> ing the 1920s and 1930s in terms of R&D competition. That is, the
> opportunity for innovation was a function in part of the absence of
> R&D competitors in the field of synthetic chemicals and plastics
> during this period. Today many U.S. labs compete with Carbide's
> R&D lab capabilities.

The same was true in Europe where Carbide's ability to compete technologically was much weaker. The problem of stiff R&D competition was one that Versoix and Carbide's other European labs faced every day. But India was not Bayer's backyard or the home turf of other U.S.- and European-based multinationals. The same manager further noted:

> If we view our R&D possibilities in terms of product changes and
> functional use changes [see Figure 2.2], then we are limited in Eu-
> rope by our R&D size vis-a-vis our competitors to maintain a low-risk
> R&D program that concentrates on making minor changes in exist-
> ing products for existing functional uses (the X Zone).
>
> However, the acquisition of BXL changes our European R&D
> competitive ability in plastics. Here we may extend our R&D activi-
> ties to include product extensions and substitute functional uses
> (the Y Zone).
>
> In India, However, once our R&D resources are built up, the
> absence of strong local R&D competition may allow us to extend
> our R&D operations into the new product—new functional use zone.
> There are also other reasons for performing this kind of R&D in
> India.
>
> First, Indian law decrees that the Indian subsidiary spend at
> least half its earnings in India. Obviously, the entire multinational
> system would like to benefit from the Indian investment. If Indian
> earnings cannot be utilized outside India, technology produced in
> India with these earnings can be exported. Both India and the mul-
> tinational system benefit under this arrangement.
>
> Second, the cost of R&D for developing new products with
> existing, substitute, or new functional uses is expensive. It may be

FIGURE 2.2

Union Carbide: Potential R&D Capability—
Competitive Analysis

	Existing Functional Use	Substitute Functional Use	New Functional Use
Existing Product	X (Versoix)	Y (BXL)	Z
Product Extension or Modification	Y (BXL)	Y BXL)	Z
New Product	Z (New Delhi)	Z (New Delhi)	Z (New Delhi)

X = Weak R&D Resources versus Competitors
Y = Equivalent R&D Resources versus Competitors
Z = Superior R&D Resources versus Competitors

Source: Compiled by the author based on an interview with a Carbide manager.

cheaper for us to perform this work, especially the more expensive "scale up" parts in India.

Third, performing R&D work of this nature in India may be directly beneficial to the United States because of similar natural resource profiles. For instance, both nations have abundant coal supplies. The current energy dependence of the United States on oil and gas requires the development of alternative energy sources. The conversion of coal to gas is one potential alternative source of energy. But India also wants to develop this alternative energy supply and it will be much cheaper for us to do at least the expensive work there.

In 1974, the decision was made by Union Carbide to create the New Delhi lab.

3

EXXON CORPORATION:
ITS ENERGY BUSINESSES

Exxon spent $103 million for R&D in 1973 that was related to the corporation's energy businesses. These businesses were based on oil, coal, and uranium natural resources and represented about 93 percent of Exxon's worldwide sales. Related R&D outlays constituted 0.3 percent of these sales, or $26.5 billion. By 1973, 22 percent of total R&D expenditures were spent abroad. Table 3.1 presents historic data for R&D expenditures and employment over the 1965-73 period.

SUMMARIZING EXXON'S EXPERIENCE
WITH R&D ABROAD

Exxon's experience with foreign R&D activities within its energy businesses involved five foreign R&D units. All five units were created to modify technology transferred from the U.S. parent.

One was a technical assistance unit created by the English affiliate in the early 1930s to service a refining facility and modify products and located in Abingdon, England; another was a technical assistance unit created by the French affiliate in the early 1930s to service the process needs and modify products of a refining facility and located in Paris, France, and later moved to Port-Jérôme, the site of the refinery. A third was a technical assistance unit created by the Canadian affiliate in the early 1930s to service the process needs and modify products of a refining facility located in Sarnia, Canada. Still another was a technical assistance unit created by the German affiliate in the early 1930s to service the process needs and modify products of a refining facility located in

TABLE 3.1

Exxon Corporation: Geographic Distribution of Energy-Related R&D Expenditures and Professional Employment for Selected Years

Year	Total System	United States	Abroad	Percent of Total Abroad
R&D Expenditures (in millions of dollars)				
1955	n.a.	n.a.	n.a.	—
1960	n.a.	n.a.	n.a.	—
1965	70	55	15	21
1970	71	55	16	23
1972	84	63	21	25
Professional R&D Employment				
1955	n.a.	n.a.	n.a.	—
1960	n.a.	n.a.	n.a.	—
1965	1,365	1,065	300	22
1970	1,321	1,018	303	23
1972	1,369	998	371	27

n.a. = not available.
Source: Company records.

Hamburg, Germany; the fifth technical assistance unit was created by the Italian affiliate in 1965 to service the process needs and modify products of refining operations located in Fumicino, Italy.

At creation the strategic composition of R&D expenditures was 100 percent in support of existing business activities for each foreign R&D unit. Each unit was relatively small when created with only a few R&D professionals. The R&D director of each unit reported to the head of refining who reported in turn to general managers of each national subsidiary.

All five foreign R&D units continued to exist in 1974. Yet, by 1974, none existed whose primary function was to transfer technology supplied by the U.S. parent. However, all five units did experience some growth in the number of R&D professionals employed for technology transfer purposes.

The five foreign R&D units created originally as transfer technology units expanded their activities to include primarily the generation of new and improved products and processes for their national and regional markets. These indigenous technology activities constituted at least half their professional R&D personnel. For instance, an in-depth analysis of the French unit's R&D projects in energy revealed specifically that about 26 percent of the R&D professionals were still involved in technical assistance and process adaption work; 35 percent worked on product application projects primarily for Esso France; and the remaining 39 percent were involved in R&D projects performed primarily for the French and/or European region for energy conservation, environmental conservation, and new or improved products to replace products or inputs that were in short supply.

By 1974, selected characterisitcs of all five units showed that the U.K. unit was the largest R&D unit, employing 125 R&D professionals (mostly U.K. nationals). The unit's activities involved decreasing proportions of technical service work technology transferred by the U.S. parent and increasing proportions of R&D to develop new and improved technology for the European region. In this regard it was assuming increasing responsibility for regional specialization in gas, liquefied petroleum gas (LPG), aviation fuel, and engine oils.

The French unit was the smallest R&D unit in 1974 with 48 R&D professionals (all French nationals). One reason for the unit's small size was that it had been built up as a chemicals R&D facility but these activities were transferred to another R&D unit (see Exxon Chemicals analysis). Like the U.K. unit, the French unit was also performing increasing amounts of R&D to develop new and improved products for the French and European markets. These activities had increased since the energy crisis and the attending shift of some R&D professionals out of process adaption and technical assistance and into R&D projects to develop new energy conservation devices. The French unit was also responsible for specializing in industrial and process oils and asphalt for the European region.

The Canadian unit employed 68 R&D professionals (mostly Canadians) in 1973. The unit's evolution was similar to those of the U.K. and French units. The German unit employed 74 R&D professionals in 1973. Its growth included a growing amount of work to generate new and improved products for the German and European markets. The unit was responsible for specializing in middle distillates, residual fuels, and grease for the European region.

The Italian unit had grown rapidly since it was created in 1965 and employed 56 R&D professionals in 1973. The main cause of growth was Exxon Europe's desire to institute regional specialization of R&D activities among the four European R&D units. In order to achieve this goal, the Italian unit had to become larger while strengthening its technological capabilities in gasolines, automotive emissions, and wax.

The strategic composition of R&D expenditures had shifted over the years so that by 1973 approximately 10 percent of total R&D outlays was for exploratory research in support of existing business. The remaining 90 percent was R&D in support of existing business. The administrative position of the U.K., Canadian, and French R&D units changed as their directors began reporting to R&D department heads who reported directly to each subsidiary's general management. The directors of the German and Italian units continued reporting to other functional managers outside R&D. However, the geographic level of R&D responsibility had changed for the four European units as their directors also reported to a director of European R&D located in London at Exxon Europe's regional headquarters.

While each lab originated as a transfer technology unit performing technical service, all five labs felt a need to provide more sophisticated work. As one R&D director noted:

> It turned out that technical assistance was not enough to get good people and if yo· 4i·../'t offer the opportunity to do research, you didn't get the top quality people you wanted to advise your affiliates and customers on technical matters.

One result was that each affiliate began to pick up a much larger bill for particularized or, as Exxon calls it, "self" R&D (that is, R&D of interest primarily, if not exclusively, to one or at most a very few affiliates). Esso Europe had considerable say over where these "self" R&D projects were performed, since at the heart of Exxon's decision-making philosophy was the notion that "those who pay, should control." Consequently, the vast majority of this "self" R&D was performed abroad and the European R&D units grew as the European affiliates grew along with their expanding contributions to "self" R&D.

The Regional Specialization of R&D Abroad

By 1966, however, a new development began to unfold: a special category of R&D funds emerged in Europe called "European mutualized funds." In order to understand this new development, one must know that all research at Exxon was administered under one systemwide contract called the Standard Research Agreement (SRA). The presidents and executive vice presidents of each region (for example, Exxon Europe) were board members of Esso Research and Engineering, and they negotiated over R&D projects based on corporate and regional representation. Exxon's board of directors had the right of final review, and it adopted or modified the SRA based on its particular corporate view.

The actual funding and allocation of mutualized R&D funds occurred under the worldwide SRA. First, all Exxon affiliates contributed funds based on some relatively complicated rules, which related contributions to several variables. Most funds were allocated by Esso Research and Engineering to the major U.S. labs where most mutualized R&D projects were performed. A small amount of total or worldwide mutualized funds were allocated to the European labs.

However, a second pooling of funds occurred among affiliates within the European region. This second allocation was still within the SRA and accounted for most of Europe's mutualized funds obtained by the European labs. Europe, either through this second pooling or through the "self" funds, was actually funding most of its own research. For instance, when Sweden contributed to R&D, a part went to world mutualized projects (generally long-term exploratory research under control of Esso Research and Engineering); however, part of Sweden's R&D contribution also went to Exxon Europe, which supervised projects of direct interest for them. This system of European-billed R&D started approximately in 1966 when Esso Europe was created. The exact proportions of the split were not determined by Sweden but by European-based R&D people. Finally, Sweden could contract its own "self" R&D projects with any Exxon R&D unit, although the choice was generally one of the four European R&D labs.

> Apparently, smaller affiliates felt this "self" R&D need because every so often corporate officials would turn over a rock and there would be a small R&D lab. This is still probably happening today, but may be more difficult to start. These revelations created concern at corporate headquarters about R&D in Europe. The prevalent feeling was that there must be waste and duplication. [an Exxon R&D manager]

The result of corporate headquarters's concern were two in-house studies of Esso Europe's R&D activities. The first study was executed in 1964; the second was completed in 1970. The 1964 R&D study concluded that R&D should be specialized at different labs, for example, auto fuels in one lab, lubricants in another lab, and so on. However, specialization was not implemented because the existing European labs didn't want to give anything up. One Exxon manager stated frankly that, "The labs had been left to themselves for some time and now they didn't want to go the specialization route. They talked it, but didn't implement it."

Fears concerning project duplication arose again in 1970. A six-month study was commissioned to ask where European R&D had gone, where it should go, and what could be done. After reviewing the work of each European lab, the study group concluded that, overall, a large part of Europe's R&D was extremely valuable; the "self" R&D was especially getting lots of guidance and turned out to be good short-run R&D that was applied immediately; and interestingly, a small amount of exploratory research was also being performed. These latter projects were bootlegged since the labs could not get funds for them any other way. In any event, the study committee concluded that the exploratory research work was needed to maintain a competent staff and the contracting affiliates liked the quality of the long-term work. The study also discovered that the European affiliates experienced problems applying mutualized R&D to their operations. One of the study's authors later remarked:

> The philosophy was that mutualized R&D must be work that all in Exxon could use; but it was bastardized so no one used it. The para-

dox was that particularized or "self" R&D (when done well) was often picked up by other affiliates; it was easier to mutualize than the mutualized R&D.

Finally no substantial evidence was found that wasteful, duplicative R&D work was being performed by the European R&D labs. Nevertheless, the study committee felt that movement toward greater specialization was still a recommended course of action as the labs grew. Consequently a decision was made at the corporate level to specialize completely mutualized funds by particular product/technology lines. The result was that the English lab began specializing in liquified gas for motor oils; the French lab, in industrial oils, specialty oils, and asphalts; the German lab, mainly in distillates, fuels, and greases; and the Italian lab, in gasolines.

The task of implementing greater specialization started in 1971, but then came the realization, as one manager put it:

> R&D can't inflict standardization on marketing. For instance, Mercedes generally wants a particular oil or they won't approve it. Consequently, we must still do "self" R&D in Germany for motor oils even though motor oils are the province of the English lab. Also, it's hard for a newer lab like the Italian lab because the English lab, which is the oldest, has all the "specialities" and it wants to do everything and probably can do it better than the Italian lab.

Implementing greater lab specialization was further complicated because radical shifts or changes in manpower were not possible overnight. "If you reduced quickly the amount of mutualized work traditionally performed in a lab without increasing the 'self' R&D, the organizations would have lost valuable people." Exxon's senior managers felt that such cutbacks in R&D personnel could be very detrimental because their future replacement was not easy. Such recognition did not always exist. In the past, Exxon tended to let near-term business perspectives influence its commitment to R&D, especially in terms of its engineering arm. During business cycle slowdowns, employee shakeouts would occur. Corporate executives now have decided not to let R&D be cyclical. One senior manager summarized the reason for this policy:

> We used to ask ourselves, since we're not building a refinery this year, why do we need all those design engineers. Then manpower limitations would develop two or three years out in terms of having enough qualified engineers to conduct certain kinds of projects.

By 1974, lab specialization was pretty much completed in Europe. The director of the French lab felt that specialization had made each lab stronger because the best researchers in a given area of technological specialization were collected under one roof. Why was the move into technological specialization successful now and not in 1964? The director of the French lab explained:

> To implement specialization, great power was needed to tell the labs what to do. Esso Europe was filling this vacuum as the national affiliates' power was eroded by the regional command. Hard decisions were taken regarding the English lab's growth. We had to stop its growth, "make it ungrow" in terms of R&D resources and projects. A more balanced set of R&D labs was required in terms of technological skills. Ideally, we needed at least two facilities—one on the continent and one in the United Kingdom. But the reality of our situation was that we had four labs: two good medium-sized labs for product quality in France and Germany; the Italian lab was too small; the English lab was too big. The move was to make the Italian lab bigger and the English lab smaller. Yet each lab had to feel they had an assignment which if successful would insure their future existence.

AN IN-DEPTH LOOK AT THE CREATION AND EVOLUTION OF R&D IN FRANCE

The creation of R&D in France began with a small group in the Esso Standard Societe Anonyme, France (ESSAF), Refinery Department. The R&D group leader reported to the head of the refinery. There were about 40 people located near Paris around 1930. They were mainly working on adaption of refining processes for Esso France and some small amount of marketing application studies. The Port-Jérôme refinery eventually hired a professional chemist, Jacque Ballet, who built a team of about 20 researchers. In 1939 the refinery was destroyed when World War II broke out. The R&D unit was moved to Paris at that time. Their activities continued to grow to around 50 people. After World War II, the R&D group became more important with 130 to 150 people (35 professionals) but still part of refining. In 1950, the R&D unit moved back to Port-Jérôme with 100 people. (Note: Jacque Ballet went on to become president of Esso France.)

In 1954, the R&D unit had grown to such an extent that a reorganization occurred. The R&D unit became a separate department within Esso France. The lab at Port-Jérôme continued but the R&D director was now located in Paris and reported directly to the board of Esso France rather than the manager of refining. This reporting situation still existed in 1974.

In 1959, a decision was made to expand R&D activities further, especially in the field of chemicals. A new facility was constructed at Mont San Aignan near Rouen (about two and a half hours west of Paris). Operations started in 1961 when three-fourths of Port-Jérôme's lab personnel were transferred to

Mont San Aignan. The remaining quarter stayed at Port-Jérôme to continue process work in petroleum and chemicals. Approximately 27 professionals were running a large refining pilot plant in 1974 at Port-Jérôme. The plant was capable of producing up to five tons of most experimental products. While nearly all mutualized process R&D was done in the United States at Baton Rouge, the Port-Jérôme group was conducting a number of process adaption studies for petroleum refining as well as for chemicals additives work.

The Mont San Aignan lab was legally and administratively part of Esso France. The lab's director reported to a vice president of the national affiliate; however, a very strong dotted line also existed with Esso Europe Chemicals, which contracted work from the lab. The lab's personnel, including the director, were French citizens.

The Mont San Aignan lab started regional European R&D work around 1961, several years before the regional area organization was created; however, regional work greatly increased since Esso Europe was organized in 1966. The lab's managers felt that one of the big advantages of Mont San Aignan was its close proximity to the Port-Jérôme refinery (30 miles away). Each research section at the lab was responsible for some projects at the refinery's pilot plant, which forced them to go there. Representatives from each section went to Port-Jérôme on an average of twice a week. Also, Port-Jérôme oil and chemical people came to the lab for working sessions regularly. Each Friday morning one of three main R&D sections of the lab (chemical additives, industrial oils, and asphalts) had its principal researchers give a formal presentation stating what had been achieved since its last presentation (rotation period was about two months for each researcher). The people from Port-Jérôme or elsewhere who might be able to use the results were invited to these presentations. In the afternoon, project groups were formed of R&D, refining, and marketing people to decide further action on each project for the next two months.

In October-November of each year, the planning groups of Esso France met with representatives from R&D (marketing, refining, logistics, and so on). Each R&D section made proposals for next year's projects. They had to convince marketing and refining to finance those projects, which were more than technical assistance projects. A blanket budget was provided for the technical assistance projects that were not possible to define beforehand. The technical service projects versus "self" R&D projects were differentiated primarily by required funds (a $2,000 limit for technical assistance projects).

For the last several years, R&D projects have been related to business objectives. This was not always the case. Today, R&D people also promote their projects much more, by traveling to make sure implementation occurred and to drum up financing for new projects. Every day, out of 58 to 59 R&D professionals at Mont San Aignan, at least three were traveling in Europe, the United States, or other parts of the world.

The lab's performance was judged by monthly progress reports written in English and French; the number of patents developed whose value was appraised by a committee; and evaluation of projects that did not result in patents but were still considered valuable.

By 1966, the lab had reached 216 people (89 in chemicals); however, the startup (1967) of a chemicals lab in Brussels meant a continued transfer and reduction of people at Mont San Aignan. The distribution of R&D personnel in 1974 was (total employees = 189) 17.3 in exploratory R&D (chemicals); 17.9 in energy conservation (this was new); 30.9 in product shortage (this was substituting oil and chemical products or inputs for materials now in short supply); 17.9 in environmental conservation adopted for French situation; and 105 in product applications projects and process adaptation and technical assistance, of which 60 were in product work and 45 were in process and technical assistance work. The Mont San Aignan lab has three major R&D customers: Exxon Research and Engineering Petroleum (25 percent mutualized work); Exxon Chemical (25 percent mutualized work, some of which is long-range exploratory research and some marketing technical assistance); and Esso France (50 percent "self" R&D, which includes work on petroleum products produced and/or marketed by ESSAF).

Until recently, the French lab's R&D expenditures were entirely in support of existing business. This strategic composition remained largely unchanged for the petroleum business. The lab's $5.2 million budget had approximately 9 percent in long-term exploratory research (for chemicals), all of it leading to existing business activities. The remaining 91 percent was in support of existing businesses, the majority being for the petroleum business. No R&D to develop new high-risk business was performed; however, the lab's director felt this might change in the future because of the energy crisis.

The energy crisis already had caused a significant shift in R&D focus at the French lab. Before the crisis, a buyer's market existed when fuels, for instance, were in abundant supply. Under these conditions, marketing wanted R&D to develop ways to move more fuel (for example, by developing better blast furnace processes but ones that would use more fuel). After the energy crisis, the market switched to a seller's market as fuel supplies grew scarcer. Consequently, the marketing emphasis shifted to the conservation of fuel and R&D people were reassigned to new projects. Since marketing was selling everything it had, it didn't need R&D to help it move available supplies. The lab's director believed his R&D unit had the ability to generate many good new projects but the main constraint was financing, especially when market changes turned off its R&D customers (that is, marketing).

Despite the squeeze for financing, the general feeling at the lab was that the future for R&D in France was good because the French government was going to pressure Exxon to allocate more resources there and the R&D problems were often distinct from the United States because the businesses were slightly different.

The director of Mont San Aignan believed that even if the French R&D unit had not been created by the company, it would have been created after World War II. At that time, the French government decreed that any significant French industry and company in France should have R&D performed locally (that is, inside France). Since oil was listed as a "significant" industry, the Ministry of Industry's Petroleum Refining Department began overseeing technology payments. This was primarily a balance of payments function whereby the government wanted Esso France to balance payment overflows for technology against technology payment inflows for work contracted from Mont San Aignan from other national affiliates. There have been repeated meetings and correspondence with the French government on this point because outflows were much larger than inflows. This technology deficit was about $2.5 million in 1974 and was expected to grow to $4 million by 1976. This single item alone was one thousandth of France's total balance of payments deficit (about $4 billion).

The French government's policy was to reduce the technology deficits of national and foreign firms. For instance, it believed the collective impact of 100 enterprises with technology deficits similar to Exxon's would have a substantial impact on France's overall payments deficit if these technology deficits were eliminated. In Exxon's case, this policy was implemented in 1968 with a formal agreement between the government and the company's affiliate. The terms of the agreement placed a $1 million deficit limit on the French lab but on the stipulation that more R&D would be performed in France as Esso France grew. (France was the only government to the author's knowledge that checked these technology flows at this time in Europe.)

CHAPTER

4

**EXXON CHEMICAL
COMPANY**

Exxon Chemical spent $18.2 million for R&D activities in 1973. These R&D outlays were 1.2 percent of Exxon Chemical's worldwide sales. Approximately 22 percent of total R&D expenditures were spent abroad. Table 4.1 shows that foreign R&D expenditures and employment had declined in the early 1970s. However, these figures exclude technical assistance work, which comprised a large portion of the total budgets of the company's foreign R&D units.

Exxon Chemical's experience with R&D abroad involved six foreign R&D units. Five units were created to perform transfer technology work. One unit was acquired that performed indigenous technology work. However, three of the five transfer technology units were also involved in R&D projects at creation (or shortly thereafter) to develop new and improved products and processes for their respective foreign markets.

The five transfer technology units included a technical assistance unit created by the U.K. affiliate in 1955 and located in Abington, U.K., to provide market support for the affiliate's chemical businesses; a technical assistance unit created by the French affiliate in 1955 and located at Mont San Aignon, France, to provide market support for the affiliate's chemical business; a unit created by the Canadian regional affiliate to provide plant and market support for the chemicals business of its chemicals subsidiary and located in Sarnia, Canada; a technical assistance unit created by the German affiliate in 1960 and located in Hamburg, Germany, to provide market support for the affiliate's chemical business; and a regional technical assistance unit created by regional and corporate headquarters in 1965 and located in Diegem, Belgium, to consolidate market support for Exxon's chemicals business in Europe.

TABLE 4.1

Exxon Chemical Company: Geographic Distribution of R&D Expenditures and Professional Employment for Selected Years

Year	World Total	United States	Abroad	Percent of Total Abroad
R&D Expenditures				
(in millions of dollars)				
1955	5.3	5.2	0.1	2
1960	16.0	14.9	1.1	7
1965	25.8	22.6	3.2	12
1970	24.7	20.0	4.7	19
1972	17.8	13.8	4.0	23
Professional R&D Employment*				
1955	105	103	2	2
1960	254	234	25	10
1965	492	402	90	18
1970	411	316	95	23
1972	356	294	62	17

*R&D professionals are generally scientists and engineers, but figures also include anyone whom R&D directors consider a professional.

Source: Company records.

The strategic composition of R&D expenditures was predominantly R&D in support of existing business for all units. The units were all small with one to three R&D professionals (who were nationals), except the Belgian unit, which had 15 R&D professionals at creation who were from several different nations. The directors of the Canadian, English, and Belgian units reported to general managers, while the directors of French and German units reported to functional managers outside R&D.

No foreign R&D unit was created whose primary purpose was to generate new or improved products specifically for the foreign market. However, small commitments of R&D personnel did perform R&D for this purpose at the Canadian, English, and Belgian units when they were created.

One R&D unit was acquired incidentally by Exxon Chemical Company that performed primarily indigenous technology work. This was a plant and market technical service unit located in La Salle, Ontario, that was obtained when Exxon acquired a Canadian company in the early 1960s. The acquired firm was not part of Exxon Chemical Company's existing business activities nor had it received any technology from Exxon before or after the acquisition. The unit performed 100 percent R&D in support of existing business. It employed an unknown number of R&D professionals (but less than five). The unit's director reported to a functional manager outside R&D within the acquired company.

By 1974, the La Salle indigenous technology unit had not experienced much growth in the number of R&D professionals. Six R&D professionals were employed in 1974. The activities remained the same, plant and market support for businesses that were different from Exxon Chemical Company's existing businesses. The strategic composition of R&D expenditures remained 100 percent R&D in support of the affiliate's existing business. The unit's director continued reporting to a functional manager outside R&D at the subsidiary level.

Four of the original transfer technology units still existed in 1974. The German unit was disbanded in 1968 in order to consolidate chemicals R&D in Europe at the Belgian site. R&D professionals at the German unit were gradually transferred to Belgium or, if they wished, remained in Germany and were reassigned to petroleum R&D projects. After its creation in 1960, the German unit began performing a small share of exploratory research in support of existing business. However, most of the work remained technical service work for marketing (that is, approximately 90 percent of all R&D was in support of existing businesses). The unit expanded its number of R&D professionals from three to eight professionals until the process of disbandment was started. The unit's director continued reporting to a functional manager outside R&D at the subsidiary level until activities ceased.

The four surviving units had not changed their primary purpose, although all performed growing amounts of R&D work for purposes other than helping to transfer technology. All four had experienced substantial growth in personnel. For instance, the English unit employed 45 R&D professionals in 1974. Approximately 35 were engaged in technology transfer activities. R&D expenditures for these operations was 100 percent in support of existing businesses. Some 10 R&D professionals were generating technology specifically for the U.S. parent. This was primarily R&D to develop new high-risk business. The unit's director continued reporting to a general manager at the affiliate level.

The French unit employed 17 R&D professionals in 1974. Approximately 13 were performing technology transfer functions and the remaining 4 were working on projects expressly for the parent. R&D expenditures were entirely in support of existing businesses for the technology transfer functions and 100 percent R&D to develop new high-risk business for the reverse technology operations.

The Canadian unit (Sarnia) expanded its number of R&D professionals to 45 employees. Approximately 37 were employed in technology transfer activities. The strategic composition of R&D expenditures was 100 percent in support of existing businesses for technology transfer activities. The balance of R&D professionals were generating technology for the parent and for the Canadian market. This involved R&D to develop new high-risk business and a very small amount of exploratory research. The director of the unit reported to a general manager at regional headquarters.

THE BELGIAN LAB AND THE REGIONAL CONSOLIDATION OF R&D

Over time, the chemical R&D units located abroad had expanded because local management felt they needed better R&D capability. The decision for expansion was initiated at the local level, not by corporate-level management. However, in 1964, a decision was made by corporate-level management to begin concentrating chemicals R&D in Europe into one lab that would be located in Diegem, Belgium (within the Brussels metropolitan area).

> Exxon didn't want too many "splinter labs" to develop in Europe because they felt they might become too unmanageable. The main motivation for the Belgium lab was twofold: (a) a desire to centralize European R&D activity; and (b) a desire to continue tapping European technology. The latter was difficult to do from the United States because it's a lot harder to convince a French R&D whiz to move to U.S. than Belgium.
> The Japanese lab* is even a better example of this. That is, you can get someone to move from Tokyo to some other Japanese city more easily than to the United States; and we feel that as Japanese technology becomes more important, you will see more R&D done there.

The Belgium lab, called Exxon Chemical Technology, was not part of the Belgium affiliate. It was a branch or regional lab that legally and administratively was part of Exxon Chemical Europe. The decision to locate in Belgium was based on the desire to have the lab close to Exxon Chemical Europe's regional headquarters in order to improve coordination among R&D, regional managers, and chemical affiliates.

The Belgium lab started R&D activities in 1966. At first all activities supported existing business pursuits that involved technical assistance work. However, a strong exploratory group evolved as more R&D professionals joined the lab from other affiliate R&D units. Some exploratory research work was started that led directly into existing business activities. By 1972, the lab's activities were gradually redirected back into R&D in support of existing business operations.

By 1974, the Belgium lab was Exxon Chemical's largest R&D unit in Europe, employing approximately 160 people, of which 53 were considered

---

*This lab was not included in the analysis because it was part of a joint venture with two other organizations.

R&D professionals 68 were technicians, and the balance in support positions. The lab's budget was about $5.5 million. The entire amount was classified as R&D in support of existing business. About 70 percent of its operations was technical service of products and processes transferred abroad by the parent. The balance was split equally between developing new and improved products and processes for the European market and generating new technology expressly for the parent. The director of the unit reported to a general manager at regional headquarters and to a senior manager in the United States.

However, what and how much R&D the Belgian lab performed was the responsibility of each technology manager within the respective worldwide product divisions. Technical service projects were decided by marketing managers. This forced a substantial amount of interaction between marketing and technology managers and the lab's R&D management. Because the manager of the Belgian lab wanted to maintain a constant work load from year to year to ensure continuity of R&D activity, he had to remain closely appraised of technology interests in the respective product divisions.

5

INTERNATIONAL BUSINESS MACHINES

IBM spent $730 million for R&D operations in 1973. These expenditures were 6.6 percent of worldwide sales of $11 billion. Approximately $220 million, or 30 percent of total R&D expenditures, were spent abroad in 1973 (see Table 5.1).

SUMMARIZING IBM'S EXPERIENCE WITH R&D ABROAD

IBM's experience with R&D abroad involved the creation of nine R&D units. These included three transfer technology units, one corporate technology unit, and five global product units.

The three transfer technology units were all created during the 1930s and included a technical assistance unit created by the U.K. affiliate to modify adding machines and other data processing and office equipment that it manufactured; a technical assistance unit created by the French affiliate to modify adding machines and other data processing and office equipment that it manufactured; and a technical assistance unit created by the German affiliate to modify adding machines and other data processing and office equipment that it manufactured. When each unit was created, the strategic composition of R&D expenditures was 100 percent in support of existing businesses. Exact information on the number of R&D professionals was not available, however each unit was not thought to have had more than four or five R&D professionals when created. The director of each unit reported to another functional manager outside R&D at the subsidiary level.

TABLE 5.1

IBM: Geographic Distribution of R&D Expenditures
for Selected Years
(in millions of dollars)

Year	Total World	United States	Abroad	Percent of Total Abroad
1970	500	n.a.	n.a.	n.a.
1971	540	n.a.	n.a.	n.a.
1972	676	486	190	28
1973	730	510	220	30
1974	890	614	276	31

n.a. = not available.
Source: Public company records and estimates made by author.

One R&D unit was created abroad whose primary purpose was to generate technology for the parent. This was a corporate technology unit located in Zurich, Switzerland, and created by corporate headquarters in 1956. The unit was created in order to utilize outstanding European scientists to investigate the possibility of using magnetic film to replace magnetic cores in computer memories and to provide more convenient research support for growing development units in the United Kingdom, France, and Germany.

The strategic composition of R&D expenditures for the Swiss R&D unit at creation was 100 percent exploratory research in support of existing business. The unit employed 25 R&D professionals who were from several Western European countries. The unit's director reported exclusively to the head of corporate research in Yorktown, N.Y. (that is, the Swiss unit was not legally or administratively part of IBM Switzerland or the World Trade Corporation).

Five global product units were created by IBM. These included a unit located in the Netherlands and created by the World Trade Corporation and the national affiliate in 1964 to develop disc operating systems; a unit located in Vienna, Austria, and created by the World Trade Corporation and the national affiliate in 1965 to work on the theory and definition of computer languages and data processing principles; a unit located in Canada and created by the World Trade Corporation and the national affiliate in 1967 to develop software products; a unit located in Lidingo, Sweden, and created by the World Trade Corporation and the national affiliate in 1969 to develop new software programs; and a unit located in Tokyo, Japan, and created by the World Trade Corporation and the national affiliate in 1970 to develop hardware (peripheral) equipment.

The strategic composition of R&D expenditures for the five global product units was 100 percent R&D in support of existing businesses. Each unit employed approximately 15 to 20 R&D professionals when created by IBM who were all nationals of their respective nations. The director of each unit reported to the general manager of his respective national company.

All nine IBM foreign R&D units continued to exist in 1974. However, all nine had grown considerably in numbers of R&D professionals and had changed their purpose or were in the process of altering their primary R&D function. For instance, no foreign R&D unit was engaged primarily in transferring parent technology by 1974. However, seven of the nine units did have responsibility for modifying parent technology through special engineering centers that were part of these units. (The exceptions were the Austrian and Swiss units.) Also, no R&D unit abroad was developing new or improved products for its local market as its primary R&D purpose. However, at least four R&D units did allocate a minority but undisclosed amount of their R&D resources for this purpose. These were the British, French, German, and Dutch units. For instance, the British unit had developed new terminals to meet the special needs of the British banking system. Corporate R&D managers in charge of long-term R&D policy formulation expected these activities to increase in the future.

The three R&D units created originally as transfer technology units grew in size and changed the primary purpose of the R&D activities during the 1950s to indigenous technology work to develop new products for their distinctive markets. Major R&D laboratories emerged at Hursley Park, Winchester, for the U.K. unit, La Gaude, Nice, for the French unit, and Boeblingen for the German unit. However, the primary R&D purpose of these units again changed in the early 1960s to global product work as corporate headquarters began assigning R&D projects abroad to help develop the 360 line of computers.

Consequently, eight R&D units existed abroad in 1974 that were performing R&D activities primarily to develop new products for global application. Of these, the U.K., French, and German units were considerably larger than the other global product units. The former transfer technology units each employed approximately 500 R&D professionals, all of whom were nationals of their respective nations. All three units were engaged also in technology transfer activities and new product development for their particular markets. The strategic composition was 100 percent in support of existing businesses in data and word processing, except for the French and Swedish units, which performed an undisclosed amount of R&D to develop products for new high-risk business in voice communications. The directors of all these units reported to the general managers of their respective national companies. Their geographic level of R&D responsibility was global for all products and related processes developed by them.

The remaining five global product units grew in size and four of them included some technology transfer activities, but these were relatively small compared to each unit's primary activity. The Dutch unit employed approximately

250 R&D professionals, the Austrian unit 65, the Canadian unit 150, the Swedish unit 100, and the Japanese unit 150. All R&D professionals were nationals of the respective countries where their units were located. The strategic composition of R&D expenditures was 100 percent in support of existing businesses. Each unit's director reported to the general manager of his respective national company (who reported in turn to a divisional president). The primary geographic level of R&D responsibility was global for all products and related processes developed by them.

The Swiss corporate technology unit grew from 25 R&D professionals in 1956 to 70 R&D professionals in 1974. The unit's primary purpose remained the generation of technology for the U.S. parent; however, it also provided research support for the European R&D units, some of which was for developing new products specifically for the foreign market. The strategic composition of R&D expenditures remained 100 percent for exploratory research, but 20 percent now led into new high-risk business areas (in voice communications). The unit's director continued reporting to the corporate vice president of R&D in the United States. Its primary geographic area of R&D responsibility was global with secondary responsibility for the European region.

THE CREATION AND EVOLUTION OF IBM'S GLOBAL PRODUCT UNITS ABROAD

The creation of IBM's global product units abroad occurred in response to several common factors. The primary force for the development labs was an explosion of new product needs by computer customers coupled with the competitive need to improve existing data processing products and processes in the United States, Europe, and later in Canada and Japan.

Foreign government pressures or political considerations do not appear to have played a primary role in the creation of either the research lab in Zurich or the global product units abroad. The "brain drain" argument was important and was recognized, but it did not provide the primary motivation for the creation of IBM's R&D operations abroad. The primary motivation was the same as in the United States: the need for close support of existing business operations in what were rapidly becoming major market areas. However, the "brain drain" problem and attending government influence may have been contributing factors in expanding some existing development facilities, especially in France.

One might still wonder why so many labs were created in different locations instead of being centralized or at least grouped into a few technical or R&D centers? Apparently, a dual need existed: new locations had to be found for manufacturing sites; and close spatial ties had to be maintained in many cases between manufacturing and development and marketing. In fact, most of the development labs were associated with nearby manufacturing facilities. In the early 1950s, manufacturing capacities were reaching their upper limits in Endi-cott, N.Y., and Poughkeepsie, N.Y., the sites of the first labs. An earlier lab existed in Orange, N.J., dating from the 1920s, but it was disbanded and some personnel were transferred to Endicott. A decision had to be made to build new capacity in these relatively small towns or locate elsewhere. At this point, top management decided to limit IBM's work force relative to an area's total work force population by enforcing a "10 percent rule." Variations of this rule were formulated, depending on the density of a population within a 15-mile radius of possible manufacturing sites; however, the basic rule was that IBM's workforce (marketing, manufacturing, and R&D) could not comprise more than 10 percent of the total work force population.

As new manufacturing sites were selected, new development labs were con-structed in the same area. The real estate and construction division determined the locations of both manufacturing and development facilities based on several variables: labor supply, proximity to markets, environment (good schools, hous-ing), cultural attractions, recreation, and so on.

IBM's vice chairman and former chairman of the World Trade Corporation, G. E. Jones (who has been with IBM longer than any other senior manager), noted that the first foreign R&D units were created in England and France dur-ing the 1930s for mainly product modification reasons. Simple data processing equipment, principally adding machines, had to be altered to match financial and arithmetic symbols with local use. In general, the development labs have been more of an evolutionary phenomenon as opposed to the research labs. The present philosophy of the development labs was not formed at the time. They were started and evolved as service functions but were changed or done over in the early 1960s to make them administratively a part of corporate-controlled development activities. That is, development lab missions became international or corporate in nature rather than promoting exclusively the interests of a re-gion, a national company, or a unit (for example, manufacturing or marketing within a national firm under the new system). Major R&D missions were as-signed by corporate headquarters. While the performance of R&D activities be-came geographically decentralized, control over the allocation of projects and resources became much more centralized.

By 1974, each foreign development lab had worldwide responsibility for specific product-technology areas. For instance, the English lab was responsible for certain kinds of programming activities. The German lab was responsible for the "small" end of the line and minicomputer technology and mechanical print-ing technology. The Dutch lab was responsible for optical reading technology, which was originally developed there, and disc operating systems. Each develop-ment lab also had a certain number of national and regional projects in addition to its corporate missions. However, what started out, for instance, as an English or European project could have a corporate impact over time. For instance, special banking terminals first developed by the English lab for Lloyd's Bank of London were later modified for U.S. banks.

Customer technical service (product modification) was also the responsibility of five of the six European labs. Special engineering centers were located at each lab (except the Austrian lab) and worked closely with country marketing teams to solve technical problems. This work was also related to manufacturing technical service (process changes) because the labs kept engineering responsibility for products throughout their life. Consequently, the labs were responsible for implementing necessary process changes required to incorporate system or product modifications learned from customers.

All of IBM's foreign development labs are national labs in the sense that they are staffed almost completely with resident nationals. IBM feels such a policy results in a heightened sense of community. This may account for the much lower attrition rates among R&D professions in European development labs as compared to those in U.S. development labs, although IBM is not certain that other factors are not equally important in lowering the turnover of R&D personnel. The national composition of the foreign development labs may also have helped to encourage international R&D competition between labs vying for and on similar projects. The potential for competition exists because assigned lab missions often overlap one another. Such overlap is necessary because computer technology is changing so fast that broad definitions of lab projects are often required and these "big jobs" generally do not fit precisely in any one lab. This does not necessarily mean the labs are working on the same projects, per se, but that the technologies and/or project objectives may be parallel and competing.

THE ZURICH RESEARCH LAB: THE CREATION AND EVOLUTION OF A CORPORATE TECHNOLOGY UNIT ABROAD

The initial stimulus for the creation of the Zurich lab came from corporate management. The lab was established in 1956 and took possession of its present facilities in the outskirts of Zurich in 1963. Legally and administratively, the Zurich lab has never been part of IBM Switzerland or the World Trade Corporation. It has always had a corporate character.

The initial motivations for the lab's creation were: the need to support and have better contact with growing development operations in Europe and the desire to increase contacts with the European scientific community. These needs could not be satisfied adequately from the United States. The distance and traveling time factors across the Atlantic were too great, especially 20 years ago. The same motivations surrounded the creation of the San Jose Research Lab (a 10,000-person engineering and development operation existed in San Jose, along with the nearby Palo Alto-Berkeley science community).

The initial mission of the Zurich lab, as part of the research division, was to devote 100 percent of its resources (approximately 25 R&D professionals) to long-term exploratory R&D. All of this exploratory work was within existing business boundaries. The lab's primary mission was in memory technology. The objective was to develop magnetic film to replace magnetic cores. However, the capacity of magnetic cores was greatly improved by other IBM R&D units so that this mission was eventually shelved.

The Zurich lab, unlike the foreign development labs, is composed of personnel representing ten different nationalities. The majority (including the director) is European. No specialized use is made of these different nationalities regarding interaction with the different national development labs abroad (for example, a French scientist at Zurich working primarily with the Franch personnel at the development lab in La Gaude.) In terms of other operations, senior research division managers found it was not important to have a Swiss director of the Zurich lab, although it was best to have a European and one that understood the Swiss-German dialect of northern Switzerland.

They also found that the lab was able to handle comfortably 20 percent growth per year (about ten people). Beyond that, they ran short of space and good ideas given existing personnel and their inventory of ideas. Furthermore, the generation and flow of ideas required great technical vitality and a continual need for new people. Human technical obsolescence and how to regenerate the 38- to 39-year olds became major problems, just as they did in the United States.

The lab's 1974 budget ($3.5 million) was allotted among three major areas and six projects. The major areas that also formed the lab's primary units of organization were communications, solid state electronics, and physics. There were about 20 professionals and technicians permanently assigned to each major area with the rest in services and computer support.

The main difference between the early Zurich lab and the 1974 Zurich lab was the strategic composition of its work. Its entire 1974 budget ($3.5 million) was still allocated toward long-term exploratory R&D work. However, 20 percent was now being oriented toward the creation of new high-risk business systems in the voice communications field. The remaining portion of exploratory R&D was aimed at existing business problems within the data processing field.

The output of these efforts is difficult to trace because the Zurich unit shares projects with the U.S. research lab in Yorktown. Occassionally, promising projects initiated in Zurich are transferred to Yorktown because available skills and resources in the United States may enhance the probability of success at some point in a project's life. The reverse, however, is also true: a promising Yorktown project, or part of it, may be transferred to Zurich. The Zurich lab's director believes this system of shared projects has lowered the lab's "critical mass factor." That is, the Zurich lab, standing alone, would have to be a much larger operation to have a business impact. The close working association with Yorktown allows the lab to be smaller yet still make a contribution.

IBM's corporate leaders consider the Zurich lab a successful component of the research division's long-term exploratory R&D efforts. The Zurich lab's achievements support this contention.* What makes the Zurich lab especially interesting is that the successful performance of long-term exploratory R&D abroad seems to be a rare achievement. No other U.S.-based multinational enterprise is known that has experienced success with this type of foreign R&D; that is, an R&D lab established abroad to do long-term exploratory research, initiated by corporate headquarters, and reporting chiefly to corporate headquarters. Upon reviewing this situation with IBM's R&D management, including the Zurich lab's R&D managers, four reasons emerged that appeared to explain the lab's success. In no particular order of importance, these reasons were

1. The lab was not isolated from development activities abroad. Although created by corporate headquarters and not the European companies, enough development activity existed in Europe by 1960 to utilize the presence of a long-term exploratory R&D lab (for example, in voice communications).
2. The boundaries of research work were well defined. Specific missions were assigned and initially these were within the existing business interests of the enterprise.
3. The Zurich lab was part of the research division and its work was, by charter, long range and corporate managers knew this from the start. Zurich was not expected to come up with a new type of computer or typewriter technology tomorrow.
4. The Zurich lab was not isolated from domestic long-term exploratory R&D. Many of its projects have been shared. It is part of a larger long-term exploratory R&D effort. If such a pehnomenon as "critical mass" exists, it influences Zurich only insofar as it influences R&D scale economies for IBM's entire research division.

---

*Specific contributions are listed in Chapter 13, which deals with international technology transfers.

6

CPC INTERNATIONAL

CPC International allocated $13.6 million for R&D in 1973. These R&D expenditures were less than 1 percent of CPC's worldwide sales of $1,874 million. Approximately 38 percent of total R&D expenditures were allocated abroad in 1973 (see Table 6.1). CPC's experience with R&D abroad involved eight foreign R&D units. Five were created as transfer technology units, one was acquired that was an indigenous technology unit, and two were created as corporate technology units.

The five transfer technology units were created by CPC foreign affiliates during the 1940s and early 1950s to help modify product and process technology transferred by CPC from the U.S. These five units were a process technical assistance unit created in 1945 by the Belgian affiliate to service a production facility for industrial products in Vilvoorde, Belgium; a technical assistance unit created in 1950 by the U.K. affiliate to service the manufacture of industrial products and located in Manchester, England; a technical assistance unit created in the early 1950s by the French affiliate to service the manufacture of industrial and consumer products and located in Houbourdin, France; a process technical assistance unit created in the early 1950s by the German affiliate to service the production of industrial products and located in Krefeld, Germany; and a process technical assistance unit created in the early 1950s by the German affiliate to service the manufacture of industrial products located in Heilbronn, Germany.

The strategic composition of R&D was 100 percent in support of existing businesses when each unit was created. All the units employed from two to five R&D professionals at creation. These R&D professionals were all nationals of

47

TABLE 6.1

CPC International: Geographic Distribution of R&D Expenditures and Professional Employment* for Selected Years
(in millions of dollars)

Year	Total World	United States	Abroad
1966	9,355	7,245	2,110
1967	9,880	7,555	2,325
1968	10,525	7,890	2,635
1969	11,322•	8,317	3,005
1970	11,959	8,750	3,209
1971	12,282	8,962	3,320
1972	12,804	8,592	4,212
1973	13,638	8,397	5,241

*The number of R&D professionals in 1973 is as follows: United States 300, abroad 189, total 489. Data for 1966-72 is not available.

Source: Company records.

their respective countries, and they were organized so that their directors reported to the head of manufacturing or marketing within each national subsidiary.

No R&D units were created by CPC whose primary R&D purpose was to generate new and improved products expressly for a foreign market. However, one indigenous technology unit was acquired incidentally by CPC. This was a unit obtained in 1958 when CPC acquired the Swiss firm Knorr Food Products. The unit was located in Thaysen, Switzerland, and had been established by Knorr in 1940 to perform technical service to support its own manufacturing and selling functions, which it continued when CPC acquired the firm. The R&D was 100 percent in support of existing business activities. The unit had five R&D professionals when acquired (all Swiss nationals). The lab's director reported to a non-R&D functional manager.

Two foreign R&D units were created by corporate headquarters to perform R&D projects expressly for the U.S. parent. These corporate technology units were the Milan, Italy, unit created by corporate headquarters in 1960 in order to utilize outstanding scientific talent to perform long-range carbohydrate research; and the Nakatani, Japan, unit created by corporate headquarters in the early 1960s in order to utilize outstanding scientific talent to perform long-range microbiological research. The work of both corporate transfer units involved exploratory research (25 percent), R&D to develop new high-risk businesses (25 percent), and R&D in support of existing businesses (50 percent). The Italian unit employed three R&D professionals who were Italian nationals. They reported directly to senior corporate officials located in the United States. The same situation existed for the Japanese unit, which employed six R&D professionals when the unit was created.

By 1974, two of CPC's eight foreign R&D units had been disbanded. One was a technology transfer unit located in the United Kingdom and the other was the Italian corporate technology unit created to generate technology for the U.S. parent. Also, several R&D units had changed or were in the process of changing their primary purpose. For instance, two foreign R&D units that were created originally to perform primarily transfer technology work had evolved to the point that their R&D activities were geared primarily toward the development of new or improved products and processes expressly for the European markets. These units were the Belgian unit and the German unit at Heilbronn.

The Belgian unit had expanded operations to include 55 R&D professionals by 1974. This growth began in 1965 when European managers decided to concentrate R&D resources at the Belgian unit's site in order to make the unit an R&D center for CPC's industrial products business. The strategic composition of R&D experienced a shift at the Belgian unit into exploratory research leading into existing business activities. Approximately one third of total R&D expenditures were supporting exploratory research, while the remaining two thirds supported existing business operations. The director of the Belgian unit began reporting directly to a European R&D director in 1967 when CPC Europe was organized as a regional division. The unit's geographic level of R&D responsibility also expanded to embrace the European region.

The German unit at Heilbronn had expanded R&D operations to include 75 R&D professionals by 1974. This growth occurred because the managers of the German subsidiary decided in the early 1960s to concentrate several technical service teams based at different German plants at one plant site (Heilbronn) where lab and pilot plant facilities existed; and European managers later decided to make the German unit at Heilbronn a regional R&D center for its consumer business. By 1974, the strategic composition of R&D expenditures was one third exploratory research and two thirds R&D in support of existing business operations. The director of the unit reported to CPC's European R&D director, while the unit's geographic level of R&D responsibility included the European region.

Three of the five R&D units created originally as transfer technology units continued to perform technology transfer projects as their primary function. However, one unit was disbanded in 1973 so that only two transfer technology units existed in 1974. The British transfer technology expanded from four to ten R&D professionals until the unit was disbanded in 1973. Its personnel were transferred to the Belgian R&D unit in order to concentrate further industrial products R&D. Its primary R&D purpose remained the same until it was disbanded. No change was noted in its strategic composition of R&D expenditures, its organizational position vis-a-vis other functional units, or its geographic level of R&D responsibility.

The French transfer technology unit grew in size from 4 to 10 R&D professionals, while the German unit at Krefeld expanded from 5 to 20 R&D professionals in 1974. Both units continued to perform R&D in support of local manufacturing and selling functions, although growing amounts of this work were based on technology supplied by CPC's European affiliates. The strategic composition of R&D expenditures at both units remained 100 percent in support of existing business activities. Both units continued to report to functional managers at the subsidiary level.

The Swiss unit acquired by CPC grew in size from five R&D professionals in 1958 to 25 R&D professionals in 1974. The unit's activities had also changed, now being aimed primarily at developing new and improved products and processes for consumer food and nonfood products for the European market. The strategic distribution of R&D expenditures included exploratory research (one third of the total budget) leading into existing business activities. The remaining two thirds was R&D in support of existing business operations. The unit's director now reported directly to senior board members of CPC Europe as well as to Knorr's general manager.

Both corporate technology units had experienced problems and had either been disbanded or were changing their primary purpose. The Italian unit ceased activities in 1965, five years after it was created by corporate officials. The unit did not grow in size (that is, the number of R&D professionals) or change its primary R&D purpose, which was to generate technology for the U.S. parent. The strategic composition of R&D expenditures remained entirely exploratory research. No changes occurred in the unit's organization position or in its geographic level of R&D responsibility. Overall, CPC managers felt the unit experienced no noteworthy success and when key personnel left, they were not replaced.

The Japanese unit grew from 2 to 24 R&D professionals over the 1960-74 period. However, during the early 1970s, its R&D purpose began shifting into the generation of new products and processes for CPC's Far East affiliate. By 1976, the unit's activities were forecasted to be split 50-50 between projects performed primarily for the U.S. parent and the regional affiliate. The strategic composition of R&D expenditures at the Japanese unit shifted to include more R&D in support of existing business operations (70 percent). Exploratory research accounted for 15 percent of total R&D expenditures and R&D to develop new high-risk possibilities represented the remaining 15 percent.

R&D spending at CPC is divided between its industrial and consumer businesses, both in the United States and abroad.* For the industrial business,

---

*Each European affiliate pays a fixed percentage (less than 1 percent of sales) to support European regional R&D. No additional funds are provided by the U.S. parent.

company officials noted that every major national company with a corn-grinding plant had a small, quasi-R&D unit. Some of these were units doing long-range plant process assistance work for production. Others were involved primarily in quality control but became involved in minor product change work. Over time, some of these units were consolidated into country or affiliate R&D units. A senior R&D manager in Europe explained this process:

> I entered the firm through the German affiliate in 1961. The "so-called R&D" was part of quality control . . . one or two professional chemists. Wherever the German company had a plant, this was the case. My job was to centralize these scattered "R&D activities." The German lab [Heilbronn] was started in 1963 with about 40 people with pilot plant activities plus two floors of labs. I reported to a new V.P. of R&D for both industrial and consumer food who reported to the country manager. R&D work was limited because we were mainly concerned with product extensions. Later, new product development was started in consumer foods. Most industrial technology came directly from the United States; but this changed during the 1960s.

European regional consolidation stimulated this process. When CPC Europe was formed in 1967, the different country plants were specialized along particular product lines, for example, the French plant specialized in corn starches. A technical director for Europe was appointed and began to consolidate and specialize affiliate R&D activities from Brussels. (The British unit was later moved entirely to Belgium as part of this consolidation drive.) As part of this specialization and consolidation drive, three CPC Europe R&D labs now work on R&D projects of interest to at least more than one European affiliate. The three labs are the Heilbronn, Germany, lab; the Thayngen, Switzerland, lab; and the Vilvoorde, Belgium, lab.

The first two labs work only on consumer business projects, while the Belgium lab is devoted to industrial business R&D work. Data collected for these three labs at the time they were created versus their status at the end of 1973 showed a shift in the original purposes of the labs away from local affiliate interests to "European" projects of overriding view by 1970; and the performance of a growing amount of long-term exploratory research leading into existing business areas. The long-term exploratory research work in Europe includes over 20 projects to develop new product and new process technology. Some of the new product work for industry is linked to potential new process technology developed by other industry customers. For instance, future process technology developed by the candy industry may influence existing intermediate products supplied by CPC or call for entirely new products.

Consumer R&D spending has been relatively more important in Europe compared to industrial R&D spending. Exactly the reverse is true in the United

States where industrial R&D is roughly three times as large as consumer R&D. One senior CPC executive felt these inverse ratios were due to different consumer business strategies, which resulted in the buildup of a stronger consumer R&D organization in Europe.

> On the consumer side, the accent has been on acquiring products in the United States—not doing R&D internally. The best consumer R&D is done abroad in Knorr, although most of the "new product work" has been actually new formulations and so on, for product line extensions.

The main technologies for consumer foods are ingredient systems technology, dehydration technology, agglomeration-granulation technology, emulsion technology, and packaging technology. The emergence of ingredient systems technology alone indicates the growing role of science (chemistry in this instance) in the food processing business. Because tastes varied for Knorr soups across European countries, the development of product formulations matching local tastes was a critical success for marketing. But, before 1965, product formulation was a craft of chefs. Product developments came from them. The idea to move from cooks to chemists gained acceptance about this time.

7

OTIS ELEVATOR
COMPANY

Otis Elevator Company spent $13.6 million for R&D in 1974. These R&D outlays were 1.5 percent of worldwide sales of $967 million. However, the data in Table 7.1 show that R&D expenditures increased substantially abroad along with the emergence of corporate-level R&D operations in the United States during 1974. While the company's R&D expenditures were small compared to some multinational enterprises of similar size, a large share (45 percent) of total R&D expenditures was allocated abroad.

SUMMARIZING OTIS ELEVATOR'S EXPERIENCE WITH R&D ABROAD

Otis Elevator's experience with R&D abroad involved the establishment of nine foreign R&D units. These included four units created to perform transfer technology work, one unit created to perform indigenous technology work to develop new products and processes expressly for the European market, and four units acquired abroad that performed indigenous technology work, mainly technical support of local operations.

The four transfer technology units created by Otis were a technical assistance unit created by the German affiliate in the early 1950s to modify products manufactured for the German market by the affiliate; a technical assistance unit created by the U.K. affiliate in the early 1950s to modify products manufactured by the U.K. affiliate for the U.K. market; a technical unit created by the French affiliate at Argenteuil in the early 1960s to modify products manufactured by the French affiliate for the French market; and a technical assistance

TABLE 7.1

Otis Elevator: Estimated R&D Expenditures for
Principal Business and Geographic Areas
for 1973 and 1974
(in millions of dollars)

	1973	1974
Corporate R&D*	0	1.0
North American elevator-escalator operations	2.5	2.5
International elevator-escalator operations (all in Europe)	2.5	5.3
Diversified operations	3.8	4.8
United States	3.1	4.0
Abroad (all in Europe)	0.7	.8
Total system	8.8	13.6
Percent abroad	36	45

*Not created until 1974.
Source: Company records.

unit created by the Italian affiliate in the mid-1960s to modify products manufactured by the Italian affiliate.

When each unit was created, the strategic composition of R&D expenditures was 100 percent in support of existing businesses in elevators and escalators. Each unit was started by shifting one or two engineers from contract engineering into product modification work. These engineers were all nationals of
their respective countries and they reported to the head of the engineering department of each subsidiary.

Five R&D units performed indigenous technology work when they were
created or acquired by Otis. One indigenous technology unit was created that
performed R&D activities primarily to develop new and improved products for
the foreign market. This unit was the European headquarters unit created by the
European regional affiliate in 1972 and located in Paris, France, to perform and coordinate R&D operations in support of the European elevator-escalator business.
The strategic composition of R&D expenditures for the European headquarters
unit at creation was 100 percent R&D in support of existing business operations.
The unit's size was small, composed only of a few R&D professionals, but its
function was to coordinate the activities of other R&D units as well as to per-

form R&D itself. The unit's director reported directly to the president of Otis Europe.

The four indigenous transfer units acquired by Otis Elevator were a technical assistance unit acquired incidentally in the 1950s from an Austrian company that was performing R&D in support of its local elevator business; a technical assistance unit acquired incidentally in 1964 from a Spanish company located in San Sebastian that was performing R&D in support of its local elevator business; a technical assistance unit acquired incidentally in 1973 from a Spanish company located in Madrid that was performing R&D in support of its local elevator business; and an R&D unit acquired incidentally in 1970 and located in France and performing R&D in support of its local business in horizontal transportation products and systems (part of Otis Elevator's diversified operations division).

All four acquired indigenous transfer units were improving products for their respective markets except the Saxby unit, which was also developing new products expressly for the subsidiary's existing businesses. The work of all four units did not involve technology supplied by Otis Elevator; consequently, the units were classified as indigenous technology units when they joined the Otis Elevator system.

The strategic compositions of R&D expenditures for the Austrian and two Spanish units were 100 percent in support of existing business in elevators when they were acquired by Otis Elevator. Each unit had only one or two R&D professionals who were nationals of their respective countries. They reported directly to the president of each acquired company. At the time of acquisition, the strategic composition of R&D expenditures for the Saxby unit was primarily in support of existing business operations in railway signalization with a small amount to develop new high-risk business in baggage transport equipment. The unit employed six R&D professionals, all French nationals, and the unit's director reported directly to Saxby's general manager.

All nine foreign R&D units created or acquired over the 1950-73 period continued to exist and perform R&D activities in 1974. The four transfer technology units created by Otis had changed their primary R&D purpose by 1974 from primarily transfer technology work to primarily indigenous technology work. All four units created originally as transfer technology units continued to perform some transfer technology activities, though by 1974 the majority of these were intra-European transfers rather than from the U.S. parent. All four units expanded their foreign R&D purposes by adding R&D activities to develop new products for the European region. Approximately one-half of each unit performed projects for European application. The German and French (Argenteuil) units increased the number of R&D professionals to 25 at each unit by 1973. One reason for their increase in size was that both units obtained additional personnel when their subsidiaries merged with other foreign firms that had their own R&D units.

R&D expenditures for all four units were classified as 100 percent in support of existing businesses in 1973. The director of each unit continued reporting to the head of engineering within their subsidiary for local projects. However, each unit's director also started reporting to the director of the European headquarters unit for projects funded by the latter.

The four transfer technology units created by Otis Elevator evolved into indigenous R&D work when the French and German units began supplying technology in elevators and escalators during the 1960s for their respective affiliates, especially for the small elevator business. As this business evolved, the U.K. and Italian units were assigned R&D projects to generate new products and production technology for the small elevator business in the late 1960s by European headquarters.

The indigenous technology unit created abroad by Otis (the European headquarters unit) had not changed appreciably in size since it was created. Its function was the same, the performance and coordination of R&D designed to develop new products and components for European elevator operations. Nevertheless, it is misleading to view this unit in isolation since it had grown by tapping the resources of the four units that had evolved into indigenous technology work.

The R&D expenditures allocated for indigenous R&D activities for all five units (about $3 million) were classified by Otis Elevator managers as 100 percent in support of existing businesses in 1974. However, earlier expenditures in the late 1960s and early 1970s were R&D to develop new high-risk business because Otis Europe did not have a backlog of production, marketing, and technology experience in the small elevator business.

Three of the four indigenous transfer units acquired by Otis (the Austrian and two Spanish units) had not increased their number of R&D professionals by 1974. Their foreign R&D purpose remained oriented toward the performance of technical service and product modifications work for the local market. The strategic composition of R&D expenditures was unchanged, remaining 100 percent in support of existing business activities. R&D directors reported to their companies' general managers. Only the indigenous technology unit acquired from Saxby grew in size, at about a 20 percent rate, since it was acquired by Otis Elevator in 1970. It presently employs 12 R&D professionals. No other change was noted by Otis Elevator executives.

THE EMERGENCE OF R&D IN EUROPE

Historically, Otis's main impact has been in the area of large elevators for high-rise buildings. As long as this product-market strategy prevailed, R&D activities were destined to be limited in the United States and abroad for three reasons.

First, large-scale R&D resources for long-term work were not allocated because the future market for large high-rise buildings was uncertain, being a function of the construction industry, which itself is cyclical and uncertain. Second, the potential elevator-escalator market for high-rise buildings even under the most optimistic forecasts was never sufficiently large to justify heavy R&D expenditures. The entire vertical transportation world market was only $3 billion in 1974. Third, and perhaps most important, Otis's product-market emphasis led to a limitation of R&D activities by the dominance of short-term engineering needs over longer term R&D activities. Specifically, the construction of specialized elevator equipment for one-of-a-kind buildings called for a job shop production process that kept the firm's engineering, and hence its R&D attention, riveted to short-term development work at the expense of longer term R&D work. Over the years, this attention span was limited by the expediencies of business to vertical transport problems presented by the current or upcoming major contracts. The internal transportation system of each major new building was, in a sense, a new product development by virtue of different building design and different traffic patterns imposed by the building's location. The pressure on engineering to complete these contracts kept its R&D service units from developing its own longer range programs around which to lobby for greater R&D funds.

Also, U.S. labor union control inhibited standardization of products and encouraged a job shop production process. By legal contract, elevator assembly had to occur at the building site rather than at the factory. However, the absence of labor union restrictions in Europe together with different product-market requirements led to a different production process and new role for R&D abroad.

The existence of a growing high-rise market in Europe allowed the principal European elevator-escalator affiliates to develop considerable experience in contract engineering for specialized jobs, especially after World War II. In 1960, Otis Germany was permitted to design its own escalator for the German market. This decision represented a complete turnaround from the policy that Otis U.S. would supply all product technology. (Subsequently, both escalator product and process technology were transferred back to the United States.)

In the same year, engineering groups along with marketing units of different European affiliates began to pressure Otis U.S. for more freedom to develop their own technology in order to enter the small elevator-apartment market. Otis was not in this market in the United States and consequently Otis U.S. technology was not available for Otis Europe to enter this European market.

In 1960, the president of Otis France successfully championed the notion of a standardized or modelized small elevator for Europe.

Otis executives felt that rationalization of production by national specialization lowered the opportunity to make contract-by-contract innovative changes under job shop situations. However, little more than the most expedient type of

R&D work had been done under job shop conditions and European R&D managers felt a need for longer term R&D work. They had no measurement of technological output over the last decade but they believed the company's development of new elevator-escalator component-product output had fallen below a level necessary to meet their needs.

The European view gained greater support as the European Common Market gained strength and, by 1964, Otis decided to implement the new strategy. A merger and acquisition program was launched throughout Europe in order to expand operating facilities quickly. Large factories were either built, acquired, or expanded in Germany, France, and Italy and rationalized so that the product line was not produced completely in any one country.

By 1965, the elements of European manufacturing rationalization programs began to be introduced around standardized elevator models. A European regional headquarters was established and Joris Schroder was instructed to establish an R&D group to coordinate "European" R&D projects. By 1970, general and specific R&D assignments were determined by Schroder for the various European affiliates. Each national engineering department received a specialized R&D mission. For instance, the French specialized in materials, especially steel sheeting; Italy in medium-sized and small mechanical engines; Germany in electronic and medium-sized mechanical machines; and the United Kingdom in high-speed machines and related components.

The process of national R&D specialization is a long-term goal and is being implemented as various departments grow rather than by cutbacks. Nevertheless, a number of exceptions creep in and Otis Europe has found it cannot be wedded to strict specialization. Questions of motivation have meant allowing some duplication of general R&D activity.

II

AGGREGATE ANALYSIS OF THE ESTABLISHMENT AND EVOLUTION OF R&D INVESTMENTS MADE BY SEVEN U.S. MULTINATIONALS

8

R&D CREATION ABROAD

Between 1931 and 1974, the seven U.S. organizations in this study created 42 R&D units in 14 foreign countries. In addition, 5 of the 7 organizations also acquired 13 foreign R&D units. These acquired R&D units are analyzed separately in Chapter 10 because they were all incidental acquisitions; that is, each was part of a total business that was purchased primarily to acquire assets other than the R&D facilities.

In almost every case, the seven parent organizations created R&D units abroad for reasons directly related to the performance of the R&D function. Non-R&D goals, such as monitoring foreign R&D activities, taking advantage of "cheap" R&D labor, or using "trapped" or "blocked" funds, played almost no part in the investment decisions.

The 42 R&D units created abroad fell into four categories, determined by the R&D function they performed:

1. Transfer Technology Units (TTUs). Thirty-one R&D units were created to provide technical services to the parent companies' manufacturing subsidiaries and their customers, after it had become evident that the companies' U.S.-based R&D units were having difficulty in transferring technology because of language differences, distance, and related inconveniences.
2. Indigenous Technology Units (ITUs). Two R&D units were created to develop new and improved products expressly for foreign markets. These products were not the direct result of new technology supplied by the parent organizations.
3. Global Product Units (GPUs). Five R&D units were created to develop new products for worldwide production.

4. Corporate Technology Units (CTUs). Four R&D units were created to generate new technology expressly for the parent corporation in the United States.

Certain hypotheses have been developed to explain the process leading to the formation of these four particular types of R&D units. These hypotheses are based on several sources of information: the observations of managers interviewed on the subject of R&D unit information, quantitative information on selected characteristics of foreign R&D units, and relevant empirical and theoretical studies. Together, these sources of information indicate that the purpose for which an R&D unit was established affected its size, its location, the timing of its formation, and its geographic and administrative ties to other organizational units within the organization.

TRANSFER TECHNOLOGY UNITS

A necessary condition for the formation of a transfer technology unit in a foreign country was the existence of a manufacturing subsidiary. At the time of their formation, all 31 such units were administratively linked to a manufacturing affiliate of the parent corporation. Technical service units were often located within manufacturing plants, or a short distance away, because the product changes requested by customers usually involved some slight alterations in manufacturing processes. To effect these changes, several groups—R&D, marketing, engineering, and manufacturing personnel—usually found it necessary to communicate frequently, preferably face-to-face. Table 8.1 shows that the majority of TTU's (84 percent) were located at the manufacturing site.

TABLE 8.1

Location of Technology Transfer Units in Relation to Other Organizational Units

Location	Number of Units	Percent*
Manufacturing sites	26	84
National headquarters	3	10
Regional headquarters	2	6
Total	31	100

*All percents rounded.

Although 74 percent of the R&D units in this study were TTUs, they were formed in only 9 of 40 nations in which the 7 parent companies had established production operations. Several factors, described later in this section, suggest that companies were not inhibited from forming TTUs by the cost of these operations. Manpower needs were relatively small and setup costs were low. The characteristic that determined whether or not a transfer technology unit should be established seemed to be the nature of the product or process technology, that is, whether standardized or unstandardized.

A number of the managers interviewed observed that businesses based on readily standardized products or processes did not need transfer technology units. Most of Corning Glass's business, for example, was based on standardized products and processes that required little modification once the company's R&D facility in Corning, N.Y., had solved the initial technical problems. Although Corning had more than 70 plants throughout the world, the company did not create any R&D units outside the Corning area, except in its electronics-related businesses.

Additional support for the hypothesis that TTUs were established where products or processes were not yet standardized comes from two types of data: the geographic location of transfer technology units (by country) and the date of unit formation. Table 8.2 shows that the majority of TTUs were located in three European nations and in Canada. Nineteen (61 percent) were located in Germany, Great Britain, and France. The units in these countries, plus the

TABLE 8.2

Location of Transfer Technology Units by Country

Country	Number of Units	Percent*	Cumulative Percent
Great Britain	7	23	23
Germany	6	19	42
France	6	19	61
Italy	2	7	68
Switzerland	1	3	71
Belgium	2	7	77
Netherlands	1	3	80
Canada	4	13	93
India	2	7	100
Total	31	100	

*All percents rounded.
Source: Company records.

Canadian units, accounted for almost three-quarters of the total. In all four of these countries, the parent companies made early manufacturing investments for the products associated with the R&D function, and most of the transfer technology units were established early in the foreign investment process. The average formation date for technology transfer units was 1953.

Table 8.3 shows that 52 percent of the units were established before 1955. The average year of TTU formation in Great Britain, Germany, and France was 1948. These three countries accounted for 85 percent of all TTUs established before 1955. For other countries, the average date of establishment was 1960.

TABLE 8.3

Distribution of Transfer Technology Units by Year of Formation

Year of Formation	Number of Units	Percent*	Cumulative Percent
1931	4	13	13
1933	3	10	23
1945	1	3	26
1950	1	3	29
1951	5	16	45
1955	2	7	52
1958	1	3	55
1959	1	3	58
1960	1	3	61
1961	3	10	71
1963	2	7	78
1965	3	10	88
1967	1	3	91
1971	1	3	94
1973	1	3	97
1974	1	3	100
Total	31	100	

*All percents rounded.
Source: Company records and data gathered at interviews.

Conversely, several managers noted that manufacturing subsidiaries in Latin America, Africa, and the Far East—all of which had smaller markets than the three European nations and Canada—did not require permanent R&D units to perform technology transfer services. Product and/or process technologies had been standardized by the time manufacturing investments were made in these countries. When technical problems arose, they were handled by temporary R&D teams from the United States or Europe.

In the few instances where transfer technology units were created outside the major foreign markets in Europe and Canada, each of the countries involved had the potential for a large market with unique characteristics (for example, Union Carbide's investments in agricultural chemicals in India). Also, most TTUs in non-European nations were established after 1965 since these markets were developed later than their European and Canadian counterparts.[1]

With regard to the conditions that determined whether or not a transfer technology unit would be formed, several managers suggested that there was a second consideration beyond standardized versus unstandardized technology. Management had to feel certain that there would be a continuing rather than a sporadic need for technical services before recommending that a permanent organizational entity should be established.

Once the need was recognized, a permanent organization might be formed with a staff of only one or two professionals. Table 8.4 shows that 32 percent of

TABLE 8.4

Number of R&D Professionals in Newly Formed Transfer Technology Units

Total R&D Professionals	Number of Units	Percent*	Cumulative Percent
One	1	3	3
Two	10	32	35
Three	3	10	45
Four	5	16	61
Five	2	7	68
Six	6	19	87
Seven	1	3	90
Eight	2	7	97
Fifteen	1	3	100
Total	31	100	

*All percentages rounded.
Source: Company records and data gathered at interviews.

the units started out with two professionals. The majority (87 percent) had six or fewer R&D professionals. The managers interviewed stated that the number of professional staff was determined by the expected frequency of technical assistance projects. Several managers also mentioned that the size of the parent company's initial R&D investment was relatively low because so few people were needed to initiate and carry on technology transfer projects. Interviewees noted that a single R&D professional could handle several projects simultaneously.

Since the R&D work in a TTU usually involved only minor changes in existing technology, technical risks were low. Market risks were also low because R&D project investments were, in most cases, less than $3,000, far below the anticipated return from incremental sales.[2]

There were several reasons for the low initial investments of the TTUs in this study. Existing facilities associated with other direct investment activity could be used for R&D activity, and managers of other manufacturing functions could oversee the R&D units. It was seldom necessary to hire U.S. or other foreign professionals with specialized skills or a broad range of skills. Officials at Otis Elevator, for example, stated that they had staffed all their transfer technology units by shifting a few contract engineers into technical service work. These observations were supported by the data, which showed that 29 of the 31 transfer technology units (94 percent) employed only host national R&D professionals when they were established. Some managers noted that, whenever necessary, a specialist could be transferred to the TTU on a temporary basis.

Several managers stated that functional managers (other than R&D managers) were put in charge of technical operations because they had more relevant experience with regard to the practical application of technological changes in products and processes. They could more accurately determine which product or process changes they or their customers required to improve sales and/or reduce costs. The data reinforced this argument. It showed that in 28 of 31 cases, transfer technology units were "subordinate" at the time of their creation to other, non-R&D functional units in the national subsidiary, that is, the director of the TTU reported directly to a manager of another non-R&D functional unit rather than to a general manager or another R&D manager at the national level.

The data also indicated that the 28 subordinate TTUs performed R&D primarily for the countries in which they were located. This inference was supported by data that showed that the market size that supported these R&D investments was national rather than regional. Specifically, the 28 subordinate transfer technology units had national R&D responsibility, that is, they performed R&D services only for customers within the nation in which they were located. There were three exceptions, however: Union Carbide's Swiss unit and Exxon Chemical's Belgian and Canadian units. These three R&D units were created by regional headquarters to help transfer U.S. technology through technical service and product-process modification of chemicals and plastics for several national markets. (Note: Canada is organized as a regional division by Exxon.) All three R&D units employed foreign nationals, occupied separate facilities, and, on the average, used more R&D professionals when they initiated operations than those transfer technology units that were created to serve national markets.[3]

INDIGENOUS TECHNOLOGY UNITS

Regional managers for two U.S.-based multinational organizations (Otis Elevator and Corning Glass) created two ITUs in Europe to develop new and improved products expressly for the foreign markets in which the ITUs were located. Each of the organizations that created a single foreign indigenous technology unit had already made substantial foreign manufacturing investments, but had not created other ITUs abroad. One might thus infer that factors other than simply the existence of foreign manufacturing activity played a role in the decision to create indigenous technology units.

The decision process that resulted in the formation of ITUs by Corning and Otis incorporated three events. First, managers in a foreign country perceived that the current stream of new and improved products developed in the United States was (or would be) insufficient for the needs of their foreign businesses. Both Otis and Corning ITU managers stated that until 1960 and 1970, respectively, new products developed expressly for the U.S. market were sufficient to sustain the growth of their companies' subsidiaries in Europe. But after those dates, continued growth in Europe became more difficult for both organizations to maintain, as European market needs increasingly differed from the U.S. experience. As the European business expanded, European managers identified new investment opportunities that could not be exploited by using the technology developed by the parent companies. An Otis manager remarked:

> The president of the French subsidiary finally convinced corporate managers in the early 1960s that Otis was missing out on a major market by not going into the small elevator business for low-rise buildings. The main reason Otis had not entered the market was that it was not in the business in the United States and, consequently, it did not have existing product-process technology to transfer abroad. As a result, the decision was made to let Europe develop its own products and production processes for the small elevator business in Europe.

A similar situation was described by a manager who had operating responsibility for Corning in Europe:

> For a long time, the European business grew rapidly, based on products developed by Corning in the United States. We were moving along like a car speeding at 100 miles per hour. No one was interested in or even had time to be concerned with new products

designed specifically for Europe. But then managers saw growth slow down, and the car began moving along at 50 miles per hour and seemed to be ready to slow down even further. Several European managers became concerned and started suggesting that Corning Europe should begin developing its own products.

Only when managers decided that the flow of new U.S.-developed products would not be sufficient to sustain the continued growth of their businesses did they begin to implement new investment opportunities that entailed new or improved product development.[4] However, Corning and Otis managers suggested that the investment opportunities they selected had to be sufficiently different from U.S. business interests to justify forming indigenous technology units. Otherwise, the parent companies could have made a case for domestic development of new products, since appropriate R&D resources, along with potentially larger markets, would have already existed in the United States.

Finally, foreign managers noted that for a number of reasons they had to have the approval of their U.S. parent companies to create indigenous technology units. First, new product development abroad represented a distinct change in established policy. Both new and improved product development had traditionally been the responsibility of domestic R&D and engineering units. Second, when the decision involved a major product-market diversification (as it did for Otis), this had to be reviewed carefully by corporate headquarters. Third, the investments in indigenous technology units were substantial. Investment in R&D personnel was high, for example, because the commercial development of new products and processes tended to be much more complex than technical service activities. Pilot plant, tooling, manufacturing startup, and marketing of new products called for specialized skills and large-scale projects.

The decision process just described suggests that managers recommending the creation of indigenous technology units had to have political clout within their organizations. In fact, the units analyzed in this study were initiated by European regional executives and the general managers of the largest European subsidiaries. Both ITUs were located in France, a country with well-established major markets and manufacturing operations that served both France and other nations. France was also the site of each organization's regional headquarters for Europe.

Both indigenous technology units required relatively large investments, as compared, for example, to earlier investments made by Otis in transfer technology units. Managers associated with these investments at both Otis and Corning noted that ITUs required more R&D professionals than were initially involved in transfer technology units, especially when substantial process changes were involved. Professionals at Otis's indigenous technology unit mobilized R&D talent from several transfer technology units that had been formed by national affiliates in earlier years. Corning had no such professionals in Europe. Consequently, it chose to work on smaller projects for product improvement and to rely on U.S.-based R&D units to conduct larger projects, at least until the company's European unit expanded its capabilities. In both cases, managers noted that ITU projects were larger than TTU projects. Since new or improved product development required more specialized skills than did transfer technology work, the number of professionals needed for each project was higher.

The size of the initial investments in these units was also higher than for TTUs because specialists with sophisticated technical and business skills from countries other than France were needed to develop, organize, and implement projects based on new technology, or technology new to the organizations. Since new ground was being explored, the quality of the R&D professional staff was critical.

The work performed at both ITUs included R&D in support of existing businesses (requiring more time than technical service activities), R&D to develop new high-risk business, and/or exploratory research in support of existing businesses. Projects that required more R&D professionals and involved exploratory research and/or new product development were developed for a number of national markets since the technical and market risks involved were too high to be supported by a single national market. Both indigenous technology units in this study had regional R&D responsibility for projects that would be applied in a number of European nations. For Otis Elevator's new products, technical risks were high because the company had to change production technologies from custom job shopping to mass production in order to produce small elevators economically. In turn, mass production required that scale economies be realized through large output of certain elevator components whose level was beyond the consumption capabilities of any single European national market. Market risk also was high because the new products were unproved commercially and required new marketing approaches.

Several managers stated that the time required for the development of new and improved products and processes was highly variable. R&D managers said that projects often fell beyond the planning horizon of non-R&D functional groups. Thus, when non-R&D functional groups controlled new product activities, they often diverted resources to more pressing, shorter term activities involving product modification and technical service operations. One manager stated:

> The truth of the matter is that for years R&D reported to engineering and the latter was concerned with the day-to-day needs of completing important short-term contracts. Consequently, R&D people were never permitted to work on new product development because today's technology fires were too important to the head of engineering.

Once Europe's distinctive product needs were acknowledged, management created indigenous technology units with R&D directors reporting directly to regional general managers. Regional managers wanted R&D professionals to be accessible, since a considerable amount of technical information flowed from R&D personnel to the general managers who were in charge of implementing marketing and manufacturing policy for new products sold in more than one European country. As a result, the Otis unit was located within the company's regional headquarters. Corning's ITU was only an hour's drive from its regional offices in Paris, and could be reached from major manufacturing operations in 30 minutes.[5]

GLOBAL PRODUCT UNITS

One U.S.-based multinational organization, IBM, created five R&D units to develop new products for simultaneous (or nearly simultaneous) production in the United States and other foreign markets. The decision process involved two key events. The first was the decision by corporate managers to produce a world line of basic products. Corporate managers perceived that market needs for data processing in countries around the world had more in common than was originally realized. Thus, while certain elements of the new product line were being developed in the United States, other new products in the same line could be developed and manufactured abroad.

This was possible because the managers recognized that production operations of major IBM subsidiaries abroad lagged only slightly behind U.S. production. Foreign manufacturing capabilities had matured over time, while markets for new U.S. products were reached very soon after these products were initially produced in the United States. The increasing need of international product-process transfers from the United States to major affiliates abroad caused corporate managers to realize that similar transfers might occur in reverse, from foreign affiliates to U.S. operating divisions. For instance, IBM managers noted that the diffusion process had reached the point where there were virtually no exports of IBM products between the United States and major European nations.

IBM managers noted that the five global product units were not established until the 1960s, when IBM decided to produce the 360 product line for the worldwide market, rather than a line of U.S. computers, a line of French computers, and so on. Since IBM managers then decided that the 360 line should be developed as quickly as possible, for competitive reasons, all nine new computer models were introduced at the same time. Consequently, the System/360 project, one of the most ambitious and innovative ventures in commercial history, required 9 central processing models, new components technology, over 70 new peripheral machines, and completely new software. The total cost was over $5 billion, with $500 million spent on R&D alone. As T. A. Wise has written, "It was roughly as though General Motors had decided to scrap its existing make and models and offer in their place one new line of cars, covering the entire spectrum of demand, with a radically redesigned engine and an exotic fuel."[6]

The radical nature of the program spurred management to play an active role in the developing of the System/360 and subsequently the System/370. Over the 1962-70 period, the company developed large and complex sets of new products for these two systems. This meant complex R&D assignments, which, in turn, called for individuals with particular skills. Jobs became more numerous, more specialized, and less flexible.[7]

The increase in R&D positions increased setup costs for the five GPUs. Separate facilities had to be provided for each of the five units, since each was relatively large. Even so, each unit was beginning operations with the minimum number of necessary personnel and the company believed that each would need space for future expansion. All five global product units started out with 17 or 18 R&D professionals.

The five GPUs were established in the Netherlands (1964), Austria (1965), Sweden (1969), Canada (1967), and Japan (1970). Each of these nations was a major foreign market for IBM. They were concentrated in one region, Europe, along with R&D units in three other European markets, the company's largest (Germany, France, and Great Britain), which had evolved into global product work just before the creation of the five new GPUs. As the dates of formation show, IBM's GPU investments occurred during the early 1960s when the company began to develop the System/360 line.

The second critical event in the formation of a GPU was senior management's decision to assign responsibility for the initial development and production of some new products in the new line to certain foreign subsidiaries. This decision was made because the System/360 project exceeded the capacity of existing domestic development resources, not only in R&D terms but also in manufacturing, engineering, and marketing. Furthermore, IBM did not consider it practical to expand domestic resources to handle all new product work. Growth ceilings placed on major manufacturing and marketing centers in the United States prohibited such large-scale expansion of existing operations. Managers who were interviewed noted that, early in the company's history, corporate leaders had imposed ceilings on the number of IBM personnel employed in any one setting. These limitations were based on the number of manufacturing, marketing, and R&D personnel to the total work force of a surrounding community.

According to senior IBM managers, these ceilings led to the creation and expansion of new centers of operations in the United States and abroad. IBM did not think it made economic sense to start new manufacturing centers in the United States solely for product development purposes, especially since manufacturing and marketing resources already existed abroad that could be used for product development work once R&D units were created at these manufacturing and marketing sites. Conversely, new product development abroad was highly improbable until the manufacturing, engineering, and marketing functions existed abroad since U.S.-based multinationals as a group, and IBM in particular, did not make initial foreign investments to develop new products.[8]

As noted earlier, companies have found that effective product development work requires close personal interaction between R&D and other functional groups responsible for developing the new product.[9] Managers interviewed at IBM stated that the global product units were located at foreign centers of manufacturing and marketing because close administrative and geographic ties ensured efficient communications. Nearly all IBM R&D laboratories were closely associated with marketing, engineering, and manufacturing units in the United States and abroad. Four of the five GPUs were located in or near a manufacturing center. The GPU in Sweden was located at a marketing center and was responsible for software development. The products developed at the Swedish unit could be immediately transferred abroad without a manufacturing function.

These observations support the theory that the geographic location of global product units is determined by the previous existence of extensive foreign-based manufacturing and marketing investments. According to this model, major foreign manufacturing and marketing centers are usually located in large market nations, for example, the European nations, Canada, and Japan.[10] However, the decision to create a GPU outside the United States is also based on the capacity of U.S. plants to handle further activity. New product development requires a combination of marketing, manufacturing, engineering, and R&D resources.[11] If the capacity for product development in the United States is fully utilized, the company may decide that there is available capacity abroad that requires only the addition of R&D resources.

Despite the need for a close relationship between R&D and manufacturing, marketing, and engineering units, managers of new global product units were not responsible to managers of other functional units. IBM managers noted that the long-term nature of R&D work in the GPUs made it necessary for general managers of foreign subsidiaries to exercise control. These managers were closely associated with foreign manufacturing and marketing operations but were also looking at IBM's business from a long-term perspective. At the same time, the global level of this R&D responsibility required close communication between general managers of national manufacturing complexes and divisional general managers in the United States who were responsible for coordinating new product development activities at several different national locations. Consequently, all five units had independent organizational positions. R&D directors of each GPU were responsible to the general managers of the national subsidiaries in which the GPU was located.

Newly defined R&D positions at the GPUs were filled by IBM personnel and by outside professionals hired specifically for these activities. All professionals working in these R&D units were citizens of the countries in which the units were located. Although the nature of the work was specialized, it was not necessary to hire non-nationals to manage and/or to perform R&D tasks (as it was for ITUs). Professionals with business and R&D skills were found within the manufacturing and marketing units of major national subsidiaries.

One might expect that if global product projects were small, existing R&D units in the United States would have been able to expand their operations to accommodate them. The evidence shows, however, that global product units required a large investment. More R&D specialists were needed when the work involved highly complex new product forms that had to be developed in a short time. While the determinants of the costs of new product development have not been conclusively identified, available studies suggest that the principal variables are the complexity and number of products-components that must be developed for a new product or system of products and the amount of time management allocates to develop the new products.[12] According to this model, products that are large, highly complex, and developed over a short time period cost more than smaller, less complex products that are developed over a long period of time. For the five GPUs created abroad by IBM, costs were high because this work involved the initial development of complex, multicomponent product forms over a short time period.

CORPORATE TECHNOLOGY UNITS

Corporate managers of three U.S.-based multinational organizations (Union Carbide, CPC International, and IBM) created four R&D units to generate new technology for the U.S. corporate parent. The decision process followed by corporate managers had three principal events.

The first event was the decision by senior corporate executives to start or to expand corporate-sponsored exploratory research. Several interviewees noted that corporate headquarters initiated the creation of the corporate technology units when it realized that operating divisions in the United States and/or abroad did not pursue long-term R&D activities whose eventual fulfillment would occur some time after the expected tenure of the company's general managers. According to interviewees, companies did not reward operating managers for what might happen 5, 10, or 20 years in the future as a result of exploratory research performed today, and this affected the managers' choice of R&D projects.[13]

Also, managers at Union Carbide, CPC International, and IBM pointed out that corporate technology units were established during the mid-1950s and the early 1960s, when top corporate executives were worried about the long-run technological competitiveness of their organizations. This concern was more evident for the two organizations (Union Carbide and IBM) that made relatively large exploratory research commitments abroad than for CPC, which made only two small foreign R&D investments. In IBM's case, it was Thomas Watson's concern for the company's repeated noncompetitiveness in technology that led the company to establish its research division in 1956. The new division's mission was to perform all long-term R&D activities (that is, projects with sales impact five years or more in the future). This left short-term R&D operations to the operating divisions. Union Carbide also made extensive investments in exploratory research in the United States at the corporate level at the time it created a similar unit abroad. Executives thought that R&D productivity was declining over time because of an insufficient amount of exploratory research in the operating divisions.

In all four cases, the CTUs were geographically and administratively separate from other foreign investment operations because of the long-term, exploratory nature of the R & D activity. Managers preferred to preserve the independence of the CTUs in order to prevent operating units from diverting attention to shorter run problems. In each case, the R&D directors of the CTU reported to superiors at the corporate level.

A number of R&D managers noted that the charter given all four units was long term, with the goal of generating revolutionary concepts. One manager remarked, "Top corporate executives have invested in us to come up with some surprises that are relevant to the company. They don't expect us to come up with something that they can sell tomorrow." Another manager noted that "the unit's approach to major projects is to push the state of the art as far as possible, using conventional technology. However, a group is always assigned to experiment with at least one radical, nonconventional way of generating new technology." These comments were consistent with another observation about corporate technology units: all four were formed between 1956 and 1960, when corporate management in the companies in this study exhibited deep concern about the need to generate new technology. Also, the strategic composition of R&D expenditures for each CTU was oriented toward exploratory research.

The second key event in the decision to create a CTU abroad was the decision by corporate managers to employ foreign (versus domestic) scientists to generate new technology. Top managers shifted their focus from the U.S. scientific community because of scientific advances made by non-U.S. scientists that were relevant to their businesses. For instance, several managers mentioned that corporate managers in all three organizations knew of the work of foreign scientists before they invested in the corporate technology units. For example, IBM's Swiss unit was formed in 1956, shortly after several European scientists had performed work on memory cores that had a revolutionary impact on the computer industry.

The third event in the decision process was the decision to locate CTUs in an area where foreign scientists were thought to be available. Corporate executives found that they could not always persuade foreign scientists to come to the United States to perform exploratory research. Managers stated that when CTUs were large (more than 20 R&D professionals), they were located in nations like Switzerland or Belgium, countries close to other European nations where top scientists could be recruited. The two CTUs that were very small (only a few R&D professionals) were purposely located in the home countries of the leading scientists with skills relevant to the companies' needs (Italy and Japan).

But whether units were large or small, the choice of a location was based on the need to attract highly skilled professionals from a given geographic area. (It should be noted that the nations in which the CTUs were located—Italy, Belgium, Switzerland, and Japan—were not large markets for the parent organizations.)

Where and when corporate technology units were actually established is explained by a theory of the technological competition and comparative costs as these costs apply to highly skilled professional staff.[14] Geographic location is not determined by factors relevant to the foreign investment process in manufacturing. Rather, the explanation is based on the ability of organizations to use highly skilled labor in order to remain on the frontier of a technological field. According to the theory, an important aspect of competition is the investigation of new technological possibilities that may have important ramifications for future business. The implication is that the development of technology with potentially revolutionary consequences cannot always be performed entirely in the United States. Consequently, corporate managers may locate R&D investments abroad to reduce or avoid uncertainty about potential technological and/or scientific advances that may alter their businesses.[15] The reason for this uncertainty is that some foreign nations have outstanding scientists in relevant technological areas. For various reasons, these scientists are often unwilling to move for an extended period beyond national or regional borders. If U.S.-based organizations wish to employ these highly skilled professionals, they must establish R&D units outside the United States.

Finally, the size of the corporate technology units seemed to depend on the nature of the exploratory research. For instance, a few managers noted that projects involving the generation of new technology for the development of pilot products, product systems, and processes required a large number of professionals with specialized skills. But when the work involved exploratory research that was highly theoretical and did not include prototype development, it could be handled by a small number of R&D professionals.[16]

SUMMARY OF FINDINGS ABOUT R&D UNITS CREATED ABROAD

An investigation of the research and development experiences of seven U.S.-based multinational organizations provides information about why 55 R&D units were created or acquired abroad. Thirteen of the 55 R&D units were acquired when five of the seven U.S. multinationals took possession of foreign enterprises that included ongoing R&D operations. In all 13 cases, the reason for acquiring these foreign firms had nothing to do with their R&D resources. Some 42 R&D units were created outside the United States by the seven parent organizations in this study. Each of these R&D units was created abroad for one of the following purposes:

The most frequent purpose was to help in the transfer of technology from the U.S. parent to foreign subsidiaries. The evidence suggests that these transfer technology units tended to be created abroad early compared to other kinds of foreign R&D units. They were created when product-process technology was unsettled and changing to meet customer needs at a rate sufficient for managers to perceive an ongoing stream of technical service projects. These R&D units also tended to be created in higher income (European) nations. They were small investments with an average of four R&D professionals when created abroad compared to other R&D units. These R&D units performed R&D work that was entirely in support of existing businesses with primarily national R&D responsibility. They were also subordinate administratively to managers of functions other than R&D.

The least frequent purpose was to develop new and improved products expressly for the foreign market. The evidence, though very limited, indicates that two indigenous technology units were created abroad at a relatively late date compared to other foreign R&D investments. They were created when senior foreign managers perceived a need for new product development abroad because of an insufficient stream of new products coming from the parent's domestic R&D units. Both R&D units were located in France, the site of European headquarters for the two parent organizations that created them. They were rather significant investments compared to transfer technology units, involving substantial numbers of R&D professionals. They performed a broad array of R&D, including R&D to develop new high-risk business and exploratory research in support of existing businesses with R&D responsibility for the regional European market. Finally, they were administratively independent units in the sense that their R&D directors reported directly to general managers and not managers of marketing, production, or other functions.

Although given by only one organization (IBM), the second most frequent purpose was to develop new products for immediate application in both the United States and foreign markets. These global product units tended to be created at a late date relative to the startup dates of other foreign R&D units in this study. They were created when corporate managers decided to produce a single world product line and assigned responsibility for initial development and manufacture of specific new products to certain foreign subsidiaries. Global product units were created in nations that were major foreign markets and sites of production. They were large investments compared to transfer technology units, employing large numbers of R&D professionals in specialized R&D facilities. Their R&D

work was entirely in support of existing business activity when the units were created. All five units had R&D responsibility for a global market. In every case, these global product units were administratively independent of organizational units of other non-R&D functions.

Of the four kinds of R&D units created abroad, R&D units established to generate long-term technology expressly for the corporate parent were the third most frequent type. The evidence indicates that these corporate technology units tended to be created abroad when top corporate executives were concerned about the long-term technological competitiveness of their organizations and became aware of scientific advances abroad in areas related to their business activities. Corporate technology units were subsequently created in nations that made easier the recruitment of highly skilled professionals who were not interested in moving to the United States. The investment in these R&D units varied considerably, being quite small (only a few R&D professionals) when the projects were purely theoretical and quite large (more than 20 R&D professionals) when the projects involved development of prototypes. These R&D units performed mainly exploratory research to develop new high-risk business and exploratory research in support of existing businesses when they were created abroad. A small amount of R&D in support of existing business was also performed. The geographic market level of R&D responsibility was global, that is, without specific geographic boundary. Finally, all four corporate technology units were administratively independent entities when created abroad in that they reported directly to senior R&D or general managers and not to managers of functions other than R&D.

All R&D units except corporate technology units were created abroad only after the parent had established manufacturing operations outside the United States. All four types of R&D units created abroad stemmed from distinct R&D needs within the parent organization. Non-R&D factors (for example, the availability of low-cost R&D labor abroad, the need to use foreign "blocked" funds, and so on) did not play a primary role in the creation of the 42 foreign R&D units in this study, and were of secondary importance in only a few cases.

NOTES

1. These foregoing observations are consistent with studies showing that multinational manufacturing organizations make their first foreign investments in nations with large markets for products that have been sold first in the United States. For instance, see Raymond Vernon, *Sovereignty at Bay: The Multinational Spread of U.S. Enterprises* (New York: Basic Books, 1971), especially Ch. 3 on "The Manufacturing Industries" for citations of various studies.

2. By comparison, other studies indicate that work on new products involves a great deal of risk because market demand is uncertain. These studies suggest that the relationship between the purpose of TTUs and the size of initial R&D investment of foreign subsidiaries is explained by the role of scale, risk, and uncertainty factors in the R&D function. For

instance, the size of the initial investment is explained by production theory based on economies of scale, that is, few R&D professionals are needed for each project and the setup costs are low. Risk is defined in terms of size of R&D investment (small for TTUs) weighed by probabilities of technical and market success. Uncertainty is measured by the number of unknowns involved in determining the probable outcome of technology transfer services. In the case of transfer technology work, uncertainty is very low since there are rarely any unknown factors. The theoretical aspects of scale, risk, and uncertainty are covered in Raymond Vernon, "Organization as a Scale Factor in the Growth of Firms," in *Industrial Organization and Economic Development*, ed. J. W. Markham and G. F. Papanek (Boston: Houghton Mifflin, 1970). Risk and uncertainty are also covered in Edwin Mansfield, *Industrial Research and Technological Innovation* (New York: Norton, 1968), pp. 55-58.

3. The rationale for the relationship between R&D purpose and the administrative position of TTUs is supplied by studies of organizational structure and control, that is, transfer technology units were made subordinate to other functional units so that managers with responsibility for product expenditures and sales would control R&D activity. The dynamic theory of production suggests another explanation for the organizational position of TTUs: that economies are derived from experience over time. In the case of investment in TTUs, production, engineering, and marketing personnel have acquired the necessary experience with regard to particular products and processes. Thus, the administrative relationships within TTUs are determined by the need for communication between experienced functional managers for "new" R&D personnel. The pioneer article is K. Arrow, "The Economic Implications of Learning by Doing," *Review of Economic Studies* 29 (1962): 155-73. Other useful sources are F. M. Scherer, *Industrial Market Structure and Economic Performance* (Chicago: Rand McNally, 1973), p. 74; H. F. Hall, "Transfers of United States Aerospace Technology to Japan," in *The Technology Factor in International Trade*, ed. Raymond Vernon (New York: National Bureau of Economic Research, 1970), pp. 345-53; and Harry Townsend, *Scale, Innovation, Merger, and Monopoly* (London: Pergamon, 1968), pp. 4-13, in which are cited some U.K. studies of learning economies.

4. A number of other studies reinforce these observations. First, several works suggest that U.S. multinationals experience no advantage from venturing abroad solely to develop new products. See Yair Aharoni, *The Foreign Investment Decision Process* (Boston: Harvard Business School, 1966), also Vernon, *Sovereignty at Bay*, op. cit. The opportunity to develop new products for a foreign market does not usually occur unless general management, marketing, engineering, and manufacturing resources are already established in these foreign markets, supporting themselves by producing and selling products initially developed for the U.S. market. Also, empirical and theoretical studies of the growth of the foreign subsidiaries suggest that foreign managers may seek new investments in their local markets in order to insure growth. For instance, see Edith Penrose, "Foreign Investment and the Growth of the Firm," *Economic Journal* 66, no. 262 (June 1956), pp. 220-35.

5. This organizational structure is in keeping with established theories about the communication of technical information for technological innovation. Theoretical work dealing with the communication of technical information corroborates the need of foreign managers for close proximity of functional R&D operations. Since new products often require substantial changes in product characteristics, close and rapid communication among general managers and marketing engineering and manufacturing personnel is a necessary condition for indigenous technology unit activity. See Richard Rosenbloom and Francis W. Wolek, *Technology and Information Transfer* (Cambridge, Mass.: Division of Research, Harvard Business School, 1970), pp. 101-08; see also Jack Morton, *Organizing for Innovation* (New York: Random House, 1969); and Mansfield, op. cit., pp. 84-88.

6. T. A. Wise, "IBM's Big Gamble," *Fortune*, August 1965, pp. 67-72.

7. Mansfield, op. cit., p. 62.

8. Vernon, *Sovereignty at Bay*, op. cit., Chs. 2 and 3.

9. See Rosenbloom and Wolek, op. cit.; and Morton, op. cit.

10. Vernon, *Sovereignty at Bay*, op. cit., pp. 60-97.

11. See Rosenbloom and Wolek, op. cit.; and Morton, op. cit.

12. Mansfield, op. cit., pp. 62-63.

13. There are, in fact, a number of studies on the need for better communication between top corporate management and the managers of R&D projects that can have a long-term impact on business operations. For instance, see Rosenbloom and Wolek, op. cit., p. 114.

14. These ideas are summarized in Harry Johnson, "The State of Theory in Relation to Empirical Analysis," in *The Technology Factor in International Trade*, op. cit., pp. 18-19.

15. Mansfield discusses the research process as "uncertainty reduction"; see Mansfield, op. cit., pp. 57-61.

16. Ibid., pp. 62-64. This indicates that more professionals with specialized skills are needed for projects involving prototype development than for exploratory research that is purely theoretical.

CHAPTER

9

EVOLUTION OF
R&D ABROAD

This chapter investigates the causes of change in the same 42 R&D investments since they were created abroad. It looks specifically at changes experienced by two types of R&D units: transfer technology units and corporate technology units. The two indigenous technology units that were created abroad have been included in the analysis of transfer technology units, since several TTUs moved into indigenous technology activity after they were created abroad. Also, the five global product units that were created abroad are considered in connection with those transfer technology units that eventually evolved into global product activity. This is especially appropriate because all five GPUs were established by IBM, as were the three transfer technology units that eventually took on the same function.

Two factors were used to measure the direction and degree of change experienced by each R&D unit: change in purpose, for example, from technology transfer to indigenous technology; and change in the number of R&D professionals employed by the unit.

PRIOR EXPECTATIONS ABOUT THE EVOLUTION
AND GROWTH OF R&D ABROAD

There have been no empirical studies of the reasons for R&D growth, or lack of growth, abroad. However, the literature does contain some theoretical work that sheds some indirect light on the subject. In the author's judgment, the sources that provide the most plausible explanations for the evolution of R&D abroad are those that deal with the dynamic theory of the growth of the firm,

and specifically the role of scale and experience factors in the growth of an organization's units of operation.[1]

The principal notion underlying the theory of the growth of the firm as it
applies to multinational organizations is this: When there is a growth of retained
earnings at the corporate, domestic, or foreign operating level, managers in these
organizational units will attempt to reinvest these earnings in new investment opportunities for their specific organizational units.[2] Once a foreign R&D unit is
created, it can only grow if it receives more allocations to hire additional R&D
personnel. Corporate managers are more likely to make such allocations if available cash flows are increasing.

Of course, the key question remains: Given the availability of financial
resources, why do managers invest them in more R&D? One answer provided by
the literature is that more R&D is more productive than less R&D because of
scale factors in R&D operations.[3] Another response is that more R&D may also
be more productive not only when the number of R&D professionals is increased
but also as R&D professionals increase their experience and/or cumulative learning within a defined technology-business arena.[4]

The supporting evidence for these scale-related theories for R&D comes
mainly from reports of production operations showing that scale and learning
factors lower the cost per unit of output in manufacturing.[5] If we think of R&D
activities in the same terms, we can infer that the unit cost of R&D output may
fall because larger numbers of R&D professionals can overcome scale barriers
produced by the nondivisibility of R&D personnel, lumpiness of R&D project
investments, and/or the lumpiness of R&D setup costs.[6] Also, we can infer that
larger numbers of R&D professionals may speed the movement down an R&D
learning curve if these R&D professionals work on R&D projects related to each
other in a technological sense.

These theories influenced expectations for the evolution of R&D abroad in
several ways. First, it was expected that R&D units created primarily to perform
technical assistance work might grow in size as retained earnings were reinvested
by foreign subsidiaries as market size and plant capacity expanded in a particular
country or region. Greater manufacturing output would increase the need for
more R&D professionals in certain cases, because the need for technical services
for both plants and customers might also increase.

Second, it was expected that R&D units created to develop new and improved products and processes would grow larger in order to overcome scale
and/or experience barriers. These barriers seemed to be associated with larger
R&D projects involving new or improved product-process development because
the skills and knowledge needed for successful implementation became more
specialized.

Third, it was expected that R&D units created to perform technology
work specifically for the U.S. parent would also grow larger over time, for the
same reasons: to overcome scale and/or experience barriers. Consequently, it

was expected that, to realize scale and/or experience economies in R&D operations, companies would expand the size of existing R&D units rather than creating additional R&D units in the same general locations. Thus, setup costs, as well as support costs (legal, personnel, cafeteria, and so on), would be spread over a larger base. Finally, it was expected that companies would be more likely to find people inside the organization with relevant experience, specialized skills, and knowledge of the company. Consequently, the higher cost alternative, hiring people from outside the organization, would be avoided.[7]

THE EVOLUTION AND GROWTH OF TRANSFER TECHNOLOGY UNITS

All 31 transfer technology units experienced some growth in professional staff, whether or not the units changed their purpose. Managers stated that TTUs needed additional personnel because existing customers with changing product needs required technical service work. For instance, Exxon R&D managers mentioned that Mercedes-Benz requested new types of automotive oils and lubricants whenever the company changed its engine design. In addition, new customers with product needs required technical service assistance.[8]

However, several managers also mentioned that as technology transfer units matured, they experienced less growth in staff. A unit could handle a moderate number of new projects (three to five) by adding only one new R&D professional. Since R&D staff became more efficient as they accumulated learning within a relatively narrow technological field, they could handle more projects and resolve problems faster. A number of managers indicated that, in most cases, new products were introduced frequently enough to forestall reductions in the number of R&D professionals or the termination of a unit. But the introduction of new projects was not frequent enough to justify large increases in R&D personnel. In fact, three of the TTUs in this study were terminated and their activities consolidated elsewhere. However, all three units had continued to grow until they were dissolved.

The transfer technology units in this study tended to follow one of two evolutionary paths: 17 TTUs continued to perform transfer technology services in 1974 and 14 TTUs evolved into indigenous technology units. In this latter group, 11 units were still performing indigenous technology work in 1974, while 3 units (all at IBM) had changed their primary R&D purpose for a second time in the early 1960s and became global product units. (Later, as has been noted, IBM created five new global product units.)

The data on average size, average age, and average yearly growth for the 31 TTUs fall into different patterns according to the unit's changing purposes. For example, the 11 TTUs that changed their primary R&D activity to indigenous technology work were larger, older, and growing more rapidly than those TTUs that had not changed their function. Similarly, the three R&D units that evolved into global product units were considerably larger and older than the ITUs and were also adding more R&D professionals per year (see Table 9.1).

TABLE 9.1

Selected Characteristics of R&D Units Originally Created as Transfer Technology Units, Classified by Their Primary Purpose in 1974

	Still Transfer Technology Units	Now Indigenous Technology Units	Now Global Product Units
Number of R&D units (total = 31)	17	11	3
Average size in 1974 (Number of R&D professionals)	19	46	500
Average annual increase in R&D professionals*	1.17	1.87	12.1
Average date of formation	1961	1951	1930 s

*From date of creation until 1974.
Source: Company records and interviews.

The Change from Transfer Technology to Indigenous Technology

This section compares those transfer technology units that did not change their R&D purpose to those that became indigenous technology units.

Four organizations (Union Carbide, Corning Glass, CPC International, and Exxon Chemical) established the 17 TTUs that were still performing transfer technology work in 1974. These units employed a combined total of 302 R&D professionals, of whom 285 (94 percent) were engaged in transfer technology activities. Among the remaining R&D staff, nine were performing indigenous technology work and eight were performing R&D expressly for the corporate parent.

According to interviewees, these 17 TTUs had not changed their primary activities because the general managers to whom R&D directors reported had not committed the resources needed to initiate projects involving new and improved products and processes. They had not committed funds (or were only starting to do so in 1974, in four cases) because they had considered the projects too expensive and/or not vital to the continued growth of foreign operations.

However, several managers noted that, by 1974, the R&D directors of these 17 transfer technology units had expressed a desire to move into indigenous technology work. Their reasoning was similar to that of directors of TTUs that had already turned into indigenous technology units. Specifically, they were being put under mounting pressure to keep their present R&D personnel and to attract new R&D professionals. They were experiencing more pressure to keep R&D professionals, as their units' growth declined. On the surface, this seems contradictory, but one R&D director explained it in terms of his feelings about two of his project directors:

> John and Bill have become first-rate project leaders, yet they aren't sure where their present careers are taking them in terms of future professional advancement. There isn't much room for professional development in this unit and it's unlikely they will get more challenging projects unless they decide to transfer to the larger R&D centers in the United States. We could lose them.

Since the performance records of R&D directors were dependent on the performance of their best people, they were naturally interested in providing them with reasons to remain with the units. A number of interviewees stated that the best incentive they could offer their R&D people was the opportunity to become involved with more challenging technology. In short, technical service work was not complex enough to satisfy the best performers. And R&D directors believed that their units could not provide excellent technical service unless they retained their best staff.[9]

To keep their talented staff, then, directors of transfer technology units attempted to "sell" more sophisticated and expensive projects, involving product-process improvements, to managers of other functions within the subsidiaries. They were often unconvincing, however, because such projects cost more, take longer to complete, and involve more technical and market risk. They found, for instance, that a product manager was often unwilling to commit funds to a year-long project if it would consume his or her entire R&D budget for the year.

However, the dilemma of the R&D director was taken seriously by operating managers of new R&D functions because they wanted to maintain or improve the quality of technical service work. As a result, they sometimes joined the R&D director in trying to convince a general manager at the national or regional level to invest in new projects. According to R&D directors interviewed, their success or failure depended on the general manager's need for new product-process technology. If there was little need, the transfer technology unit did not move into indigenous technology work. If there was a need, the transfer technology unit became an indigenous technology unit.[10]

Managers noted that when TTUs became ITUs, they increased their professional staffs, because indigenous technology projects were larger than technical service projects. A few managers noted, however, that after their units had been involved in indigenous technology for some time, the increase in the number of R&D professionals declined. A few interviewees stated that there was a fast buildup of R&D resources when operations began at a regional level, in order to develop new or improved products and processes for a larger market. But once these resources were expanded, the increase in staff slowed down.

The data in Table 9.2 show that all those transfer technology units that moved into indigenous technology work also assumed R&D responsibility at the regional level. Managers indicated that the company needed a larger market base to underwrite new and improved product-process work.[11]

TABLE 9.2

Geographic Level of R&D Responsibility for Units Created as Technology Transfer Units, Classified by Their Primary Purpose in 1974

Geographic Level of R&D Responsibility	Number of R&D Units		
	At Date of Creation	That Remained Transfer Technology Units	That Evolved into Indigenous Technology Units
National	28	15	0
Regional	3	2	11*
Global	0	0	0
Total	31	17	11

*Does not include three IBM units that subsequently evolved into global product units.

Source: Company records and interviews.

Also, the data in Figure 9.1 reinforce the notion that change of purpose influences the level of growth in the number of R&D professionals. For instance, the distribution of R&D units shows the existence of a definite step function for the number of R&D professionals employed in 1974, when classified by the year in which the R&D units were created abroad and their 1974 purpose. The figure also reveals that all TTUs that were created before 1950 had moved into indigenous

FIGURE 9.1

The Distribution of R&D Units Identified by Their Parent Organization and Classified by Their Year of Creation Against Their Number of R&D Professionals in 1974

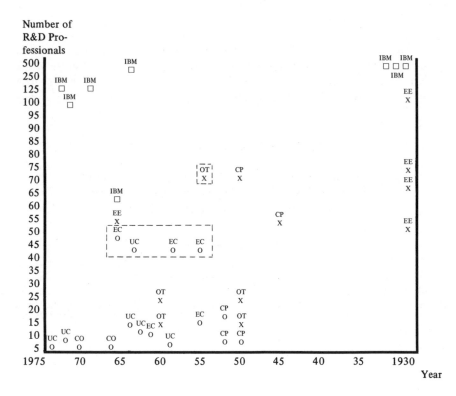

o = Purpose of R&D unit is to help transfer technology provided by U.S. parent.

X = Purpose of R&D unit is to develop new and improved products expressly for foreign market.

□ = Purpose of R&D unit is to develop new products for simultaneous production in U.S. and other nations.

UC = Union Carbide   CO = Corning Glass
OT = Otis Elevator   CP = CPC International
EE = Exxon Corp/Its Energy Businesses   EC = Exxon Chemicals
IBM = International Business Machines

Source: Compiled by the author.

technology work by 1974. The four large transfer technology units (circled in the figure) are in transition, that is, they are performing growing amounts of indigenous technology work. The four small indigenous technology units at Otis Elevator are organized for administrative purposes as one European unit with a single R&D director in Paris. They are shown as the single consolidated Otis unit in Figure 9.1.[12]

Simplifying Figure 9.1, the patterns shown in Figure 9.2 emerge once the four Otis units are consolidated into one ITU unit. This notion of a step function associated with a change of R&D purpose is also supported by the data in Table 9.3. The data show the increase in the number of R&D professionals was larger for "younger" units, that is, those units that were created as transfer technology units or that evolved into indigenous technology units after 1960. These data also support the inference that the increase in the absolute number of R&D professionals tended to decline over time until an R&D unit changed its purpose.

Analysis of the creation of R&D units abroad showed that "older" R&D units were located in large market nations.[13] Thus, the data given above would

FIGURE 9.2

Change of Purpose and the Size of R&D Investments Abroad

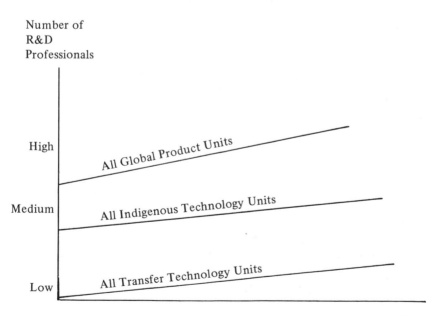

TABLE 9.3

Average Yearly Increase of R&D Professionals from Date of Transfer Technology Unit's Creation to 1974

	Remained Transfer Technology Unit	Evolved into Indigenous Technology Units*
Created before 1960		
Number of units	8	5
Average yearly increase	.98	1.7
Created after 1960		
Number of units	9	—
Average yearly increase	1.4	2.1

*Does not include three IBM units.
Source: Company records and interviews.

seem to suggest that the increase in the number of R&D professionals at older transfer technology units and indigenous technology units located in Great Britain, France, and Germany should be smaller than increases experienced by the same kinds of younger R&D units located elsewhere. In fact, the available data on geographic location show that transfer technology units located in these three large European markets have made smaller staff increases than similar R&D units located in other countries on an average annual basis (approximately same average increases as shown in Table 9.3).

The Change from Indigenous Technology Units to Global Product Units

As noted earlier, three R&D units that were originally created by IBM as transfer technology units evolved into indigenous technology units and later into global product units. The three R&D units were created as TTUs in the early 1930s, and were located in Great Britain, France, and Germany.

Managers stated that these three units did not experience large increases in R&D professionals until they started work on new product development for global application (projects associated with the System/360). However, the increase in the number of R&D professionals slowed down when the units reached capacity in facilities and in employment.

Subsequently (between 1967 and 1970), IBM created five new global product units: three in Europe, a fourth in Canada, and the fifth in Japan. The average staff size for these five younger GPUs was 143 R&D professionals in 1974; they were increasing their average size at the annual rate of 19. Unlike the three older GPUs in France, Germany, and Great Britain, the five new units were created expressly to perform specific global R&D missions within specialized technological areas. The younger global product units had to build their resources as quickly as possible, within their specialized areas, in order to compete with older global product units (as well as domestic units) that had a broader range of technological skills.

This explanation was supported by data showing that the increase in the number of R&D professionals at the five younger R&D units created as global product units was larger (24.7 R&D professionals per year for the two youngest units) than for the three older units (15.8 R&D professionals per year). While all five global product units were created with approximately the same number of R&D professionals, the difference in these staff increases suggests that when goals remained constant, increases in staff slowed down with age.

There were no available company data concerning the increase in the size of the three older R&D units after they became global product units. However, conservative estimates suggest that increases in R&D professionals since the early 1960s averaged at least 20 per year, possibly as many as 30 per year.

Managers increased the number of R&D professionals at all eight global product units because of the need to avoid project delays for a set of interrelated R&D projects that were closely tied to engineering, production, and marketing projects for the System/360 and, later, the System/370 product lines. Corporate headquarters coordinated project missions for domestic and foreign R&D units. It assigned projects for global product development on the basis of an R&D unit's relevant technological skills and availability of resources.[14]

IBM managers noted that the R&D activities of the three older global product units had begun to include more indigenous technology work when IBM altered its international marketing strategy. The company's new strategy called for greater specialization of hardware and software to meet specific user-industry needs. Because there were differences among customers in the United States and Europe (for example, U.S. versus British banking), the company had to design product systems for particular national or regional industries. As the three older GPUs began changing their purpose, between 1971 and 1974, they assigned at least 300 R&D professionals to indigenous technology work.

Some IBM managers believed that as R&D purpose evolved at the three older global product units, they would also experience an increase in the allocation of resources committed to the development of new products expressly for the foreign market. The data in Table 9.4 show an estimate of the 1974 composition of R&D professional staff, by R&D purpose.

TABLE 9.4

Number of R&D Professionals in IBM's Eight Global Product Units, 1974

	Transfer Technology Work	Indigenous Technology Work	Global Product Work
Three R&D units that evolved into global product units	150	300	1050
Five R&D units that were created as global product units	65	25	625
Total	215	325	1675

Source: Interviews and author's estimates.

THE EVOLUTION AND GROWTH OF CORPORATE TECHNOLOGY UNITS

The evolution and growth of the four corporate technology units that were created by three U.S. companies followed three of four main paths. On one path, two R&D units did not increase their number of R&D professionals or change their primary R&D purpose. These corporate technology units were Union Carbide's Belgian unit and CPC's Italian unit. Both units were eventually dissolved, one after operating 14 years and the other after 10 years.

A second path was followed by one corporate technology unit created by CPC in Japan. This unit experienced practically no increase in the number of R&D professionals assigned to perform exploratory research for the corporate parent. However, when corporate and general managers of the Far East affiliate decided to alter the R&D unit's purpose, the Japanese R&D unit began to perform indigenous technology work for the Far East affiliate in support of existing business activities. By 1974, the number of R&D professionals had grown from 6 to 23.

A third path was followed by one R&D unit (IBM in Zurich, Switzerland). This CTU experienced increases in the number of R&D professionals and some modification of purpose (though in 1974 its primary work was still in the area of corporate technology). The unit increased its number of R&D professionals at an average annual rate of 2.5. IBM managers noted, however, that part of this increase was associated with exploratory research in support of indigenous technology work undertaken by the French R&D unit during the early 1970s.

Of the four corporate technology units, only IBM's Zurich unit and CPC's Japanese unit were still operating in 1974. According to interviewees, Union Carbide's Belgian unit and CPC's Italian unit were divested because they did not produce any significant transfers of technology to the corporate parent. In fact, CPC's R&D unit in Japan did not produce any major flow of technology to the U.S. corporate parent either. Only managers of IBM's Zurich unit were able to identify important flows of technology to the U.S. parent.

Managers indicated that the poor performance of three corporate technology units was related to an insufficient allocation of R&D resources for exploratory research within a focused technological area. All three corporate technology units had small amounts of R&D resources allocated to single technological fields. This was obvious for the two CPC units, with four and six R&D professionals, respectively. It was less obvious for Union Carbide's Belgian unit; this unit employed 180 people and 37 R&D professionals. However, interviewees noted that these resources were divided among a number of unrelated or loosely related projects intended to develop different kinds of new high-risk businesses. The director of the Belgian unit confirmed that throughout its existence his unit performed only exploratory research to develop new high-risk businesses. Indeed, it may be that the Belgian unit was, in a sense, a collection of several small R&D units (unrelated project teams) within a common facility rather than an integrated operation.

By way of contrast, managers familiar with IBM's Zurich unit felt that it had survived as a corporate technology unit and increased its R&D personnel because of its larger allocation of R&D resources for exploratory research within specified technological fields that were in support of existing businesses. However, IBM managers observed that even with this focusing of resources, the Zurich unit would not have been successful had it not shared projects with the larger U.S. exploratory research unit. The general consensus was that by sharing work with the Yorktown, N.Y., research unit, the Zurich unit was able to reach the critical size needed for success.

The experience of all four corporate technology units seems to reinforce the theory that when exploratory research units are small, the probability of experiencing successful outcome is low. According to this logic, smaller R&D units face greater risks than larger units over a given time period because of the false starts, waste, and general uncertainty associated with long-term exploratory research. According to one source, the average number of successful projects for this kind of R&D activity is low (about 1 in 25 projects).[15] This does not mean that small units are by definition less productive than large units, given unlimited time. However, under time constraints, larger research teams can simply perform more projects than smaller R&D units. Consequently, it is probable that over any given period, small R&D units are less likely to hit paydirt than the larger R&D unit that is performing more projects.

RELATING EXPECTATIONS TO FINDINGS

The findings presented in this chapter conform to earlier expectations about the growth and evolution of R&D abroad, but extend beyond them due to the former paucity of information about the process of change for R&D investments made abroad by U.S.-based multinationals. These extensions are most important where the findings indicate the potential existence of three characteristics about the growth and evolution of R&D units created abroad regardless of their original purpose: large increases in the number of R&D professionals are associated with a change in R&D unit's purpose; change in purpose for these R&D units is usually in the direction of developing new and improved products expressly for the foreign market; and R&D units are more likely to be dissolved if there is no change in purpose, especially true for foreign R&D units whose original purpose is to generate technology expressly for the corporate parent.

NOTES

1. Principally, Edith Penrose, "Foreign Investment and the Growth of the Firm," *Economic Journal* 66, no. 262 (June 1956); and Raymond Vernon, "Organization as a Scale Factor in the Growth of Firms," in *Industrial Organization and Economic Development*, ed. J. W. Markham and G. F. Papanek (Boston: Houghton Mifflin, 1970).

2. Penrose, op. cit.

3. Vernon, op. cit.

4. See F. M. Scherer, *Industrial Market Structure and Economic Performance* (Chicago: Rand McNally, 1973), p. 74, for theoretical work related to economies of learning. No direct application of learning-experience concepts to the study of R&D activities is known; however, fieldwork and historic analysis of the domestic and foreign R&D experiences of the organizations in this study indicated that they were relevant.

5. See Harry Townsend, *Scale, Innovation, Merger and Monopoly* (London: Pergamon, 1968), pp. 4-9, for a concise summary of these concepts. Also F. M. Scherer, "The Determinants of Market Structure," in *Industrial Market Structure and Economic Performance*, op. cit., pp. 74-103, goes into more detail and cites a variety of studies.

6. Vernon, op. cit.

7. Townsend, op. cit.; and Vernon, ibid.

8. This increase in TTU projects is explained by studies concerned with the factors that determine U.S. foreign investment in manufacturing, that is, the development of new product technology. The underlying tenets of these hypotheses are that the product-process technologies are not standardized early in the product cycle and that they are sometimes not standardized until foreign investment occurs in small market nations, that is, late in the international product cycle. The inference of this model is that more R&D professionals will be needed at transfer technology units as manufacturing output expands in foreign markets, as long as the product-process technologies of existing products remain unstandardized or if the U.S. parent continually introduces new products whose product-process characteristics are unstandardized. See Raymond Vernon, *Sovereignty at Bay: The Multinational Spread of U.S. Enterprises* (New York: Basic Books, 1972), pp. 76-77.

9. A number of other studies emphasize that scientists and engineers need a creative environment, and place a high priority on the opportunities for achievement. For example,

see M. Scott Myers, "Who Are Your Motivated Workers?" *Harvard Business Review*, January-February, 1964, pp. 73-88.

10. A number of theories associated with the growth of multinational enterprises are consistent with this explanation. Studies show that general managers must have easy access to R&D resources in order to realize investment opportunities and that they will take advantage of these opportunities when future growth is jeopardized or in doubt. For instance, see Penrose, op. cit. Several other studies have noted a relationship between the level of R&D effort and firm size. These works are based mainly on Schumpeter's hypothesis that organizations with "market power are more apt to possess financial and organizational slack" for innovative purposes.

11. See Scherer, op. cit., p. 358, for citations of a number of studies.

12. This consolidation of the four Otis units conforms to the definition of a "foreign R&D unit" as defined in Robert C. Ronstadt, "R&D Abroad: The Creation and Evolution of Research and Development Activities of U.S.-Based Multinational Enterprises" (Ph.D. diss., Harvard University, 1975), pp. 8-9.

13. Ronstadt, op. cit.

14. Studies of technological innovation give evidence that explains the large increase in R&D professionals at global product units. The decision to expand global product units was based on the need to utilize internal engineering, manufacturing, and marketing resoures in a large-scale effort to develop complete new lines of products rapidly. As capacity was reached at existing global product units, new units were created and expanded. Here, the trade-off between R&D resources committed for product-process development versus development time appeared to be the key factor determining the level of resource commitment. For instance, Mansfield notes: ". . . the time-cost function has a negative slope, indicating that time can be decreased only by increasing total cost. As the development schedule is shortened, more tasks must be carried out concurrently rather than sequentially, and since each task provides knowledge that is useful in carrying out the others, there are more false starts and wasted designs. Also, diminishing returns set in as more and more technical workers are assigned simultaneously to the development effort." See Edwin Mansfield, *Technological Change* (New York: Norton, 1971), pp. 62-63.

15. See *Management of New Products*, 4th ed. (New York: Booz, Allen, and Hamilton, 1965), p. 9, where 52 of 55 R&D projects were found to be discarded before the development phase began.

10

EVOLUTION OF
ACQUIRED R&D UNITS

Thirteen R&D units were acquired abroad by five of the seven U.S.-based multinational organizations in this study. These five organizations and the number of foreign R&D units they acquired are shown in the following table:

Organization	Number of R&D Units
Union Carbide	5
Otis Elevator	4
Corning Glass	2
CPC International	1
Exxon Chemical Company	1

Two organizations in this study, Exxon Corporation (its energy businesses) and IBM did not acquire any R&D units abroad. One reason neither organization acquired R&D units abroad was that both followed foreign investment strategies that involved creating their own subsidiaries abroad rather than acquiring existing foreign enterprises.

However, the five organizations that acquired ongoing R&D units abroad had foreign investment strategies that included the acquisition of foreign enterprises. The 13 R&D units they obtained abroad were created originally by foreign companies that were acquired by the five U.S. organizations. However, the U.S. organizations did not acquire these companies with the expressed motive of obtaining the R&D resources of the foreign company. According to interviewees, the acquisition of all 13 R&D units was unintentional and/or incidental to other strategic purposes that motivated the purchase of the foreign firms.

Two observations support the claim of U.S. managers that these R&D units did not play any significant role in the decision to acquire the foreign firms.

First, none of the acquired R&D units represented sizable additions of R&D resources for the U.S. organizations. In fact, 11 of the 13 R&D units had fewer than 10 R&D professionals when they joined the U.S. system (all 13 R&D units had less than 20 R&D professionals). Second, no highly sophisticated or rare technological skills existed at these R&D units. All 13 R&D units performed primarily technical service work for plant and market customers that was completely in support of existing businesses. Also, 12 of 13 R&D units acquired abroad were involved in technical service work that was limited to projects for application entirely within the R&D unit's national borders.

The 13 R&D units acquired abroad performed R&D activities for two distinct purposes when they joined their respective U.S. systems. Six R&D units were acquired abroad by Union Carbide and Corning Glass that performed technical service work that was entirely transfer technology. Before these six R&D units were acquired, their primary R&D purpose was to help transfer technology from the U.S. organizations that subsequently became their parent companies. Seven R&D units were acquired abroad by Corning Glass, Otis, CPC, and the Exxon Chemical Company that were indigenous technology units. All seven R&D units performed technical service activities expressly for the foreign market that were not based on technology supplied by the subsequent U.S. parent. Tables 10.1 and 10.2 present some observations about eight factors or characteristics associated with these R&D units when they joined the U.S. system in 1974. The principal similarities and differences suggested by these tables between acquired R&D units performing transfer technology and indigenous technology are the following:

Similarities	Differences
Acquired R&D units tend to experience very low growth rates in the annual increase in R&D professions	Acquired R&D units that perform transfer technology are located in large market nations of the U.S. parent, while no locational pattern exists for indigenous technology units
Acquired R&D units tend to keep the same foreign R&D purpose they have when they join the U.S. system	Acquired R&D units that perform transfer technology tend to be bonded spatially and administratively with manufacturing facilities, while indigenous technology units are bonded with general managers of the acquired company's headquarters
Acquired R&D units tend to be composed entirely of host nationals	
Acquired R&D units tend to perform entirely R&D in support of existing businesses	
Acquired R&D units tend to maintain the same geographic level of national R&D responsibility over time	Acquired R&D units that perform transfer technology tend to be larger (twice the average size) than indigenous technology units (15 R&D professionals versus 7 R&D porfessionals for the 13 units in this study)

TABLE 10.1

Observations about Six R&D Units Acquired Abroad Performing Transfer Technology Work

Factor	When Joined System	In 1974
Date joined	Four of six units acquired since 1969	All units still exist
National location	Three of six units in the United Kingdom, one in France, one in Canada, and one in Australia	No change
Organizational location	Five of six units located at manufacturing site	No change
Total R&D professionals	Four of six units have ten to twenty R&D professionals; two others have under five R&D professionals	Five of six units have 10 to 25 R&D professionals, one other unit has 8 R&D professionals
Strategic composition of R&D expenditures	All R&D outlays are 100 percent R&D in support of existing businesses	No change; also all professionals still performing only transfer technology work
Presence of nonhost nationals at R&D units	All R&D professionals are host nationals	No change
Organizational position of R&D units vis-a-vis functional units outside R&D	Five of six units have subordinate organizational positions, that is, R&D directors report to head of functional unit outside R&D	No change
Geographic level of R&D responsibility	All six units perform R&D exclusively for customers within nation of unit's location	No change

Note: Transfer technology units are defined in this study as R&D units created abroad to help transfer technology provided by the U.S. parent by performing plant and/or customer technical service.
Source: Company records and interviews.

TABLE 10.2

Observations about Seven R&D Units Acquired Abroad Performing Indigenous Technology Work

Factor	When Joined System	In 1974
Date joined	Four of seven units acquired before 1964; other three units acquired since 1970	All units still exist
National location	No national or regional concentration; only two of seven units in one country (Spain)	No change
Organizational location	All seven units are located at headquarters of subsidiaries' general managers	No change
Total R&D Professionals	All seven units have less than seven R&D professionals; three of seven units have only one R&D professional	Five of seven units have less than seven R&D professionals; two units have 12 and 25 professionals.
Strategic composition of R&D expenditures	All seven units perform 100 percent R&D in support of existing businesses	Six of seven units perform 100 percent R&D in support of existing businesses; one unit (CPC/Knorr) performs exploratory research in support of existing business
Presence of nonhost nationals at R&D units	All R&D professionals are host nationals	All R&D professionals are host nationals except one unit with a few nonhost nationals
Organizational position of R&D units vis-a-vis functional units outside R&D	Five of seven units have independent organizational positions, that is, R&D directors report directly to general managers	All seven units are independent organizational positions
Geographic level of R&D responsibility	Six of seven units perform R&D exclusively for customers within home nations; one unit performs R&D on regional basis in Europe	Five units have national responsibility; one unit has regional responsibility; one unit has global responsibility that had regional responsibility when joined system

Note: Indigenous technology units are defined in this study as R&D units created abroad to perform R&D operations expressly for a given national or regional market that is not directly dependent upon product-process technology obtained from the U.S. parent.

Source: Company records and interviews.

These similarities and differences may, of course, break down, given not only a larger sample of acquired R&D units but one where the acquired R&D units have been part of their respective U.S. systems for longer periods of time. However, at this point, the available evidence suggests that acquired R&D units are associated with stagnant growth in the number of R&D professionals and no evolution of purpose with few exceptions.

This scenario can change with greater involvement of managers of the U.S. parent. Discussions with managers associated with acquired R&D units indicate that two patterns of involvement are followed by U.S. managers after acquired R&D units join the U.S. system. One pattern is that the acquired R&D units are simply left alone by managers of the U.S. parent. This pattern seems to happen in most cases (9 of the 13 units in this study). U.S. managers may become interested in these acquired R&D units at some later date if the R&D units become larger and more important to operations beyond the national borders of the acquired companies. The second pattern is one of intensive involvement soon after the acquired R&D units join the U.S. system. This involvement occurs when the acquisition of foreign firms represents important increases in the U.S. organization's foreign investment position (for example, Union Carbide's acquisition of BXL and Corning's acquisition of Sovirel).

The nature of the involvement of U.S. managers is one of careful review of foreign R&D operations. In Corning's case, the review resulted in a decision to create additional R&D resources abroad. In the words of one manager, "the acquisition of a number of foreign firms, including Sovirel, increased our European business very quickly. These combined firms needed a strong dose of technological resources to enable them to compete in the future." The result of involvement of U.S. managers in Union Carbide's case was not clear by 1974 since the R&D units had just been acquired in late 1973.

However, some evidence existed that suggested that the size of the acquisition activity might not be the only important variable resulting in U.S. managerial involvement with R&D units acquired abroad. For instance, acquired R&D units that received the attention of U.S. managers also had common foreign R&D purpose when they joined the U.S. system. Specifically, all four R&D units were transfer technology units. This observation suggested that prior interaction of U.S. managers with these R&D units through licensing activity might be an important variable determining the subsequent involvement in planning the future of these acquired R&D units by U.S. managers.

Through highly tentative, the principal implication of these observations is that acquired R&D units tend to be omitted by managers of the U.S. parent from R&D planning unless the acquired unit grows on its own and becomes too large and important to ignore, for example, CPC's Swiss unit (Knorr), and the chance interaction of U.S. managers through technology licensing has made them familiar with the acquired R&D unit.

Investment decisions generally take place over time and may be composed of several events and "subdecisions." Unfortunately, this nature of the decision process provides great opportunity for confusing cause and effect after an investment has been made. This problem exists for investments in R&D abroad just as it has existed for other forms of foreign direct investments. In both cases, claims have been made, and continue to be made, that noneconomic factors have caused the investment.

The confusion over the motivating factors influencing the performance of R&D abroad stems from two main sources. The first is the failure to distinguish between the decision to invest in R&D abroad versus the decision of selecting a particular national site after the initial decision is made, for example, choosing between the United Kingdom and Belgium for tax purposes after concluding that an R&D investment was needed in one nation or the other for sound R&D (economic) reasons. Second, it has been found that any discussion of non-R&D factors influencing the performance of R&D abroad must carefully distinguish between those factors causing the creation of an R&D unit; factors causing growth in the number of professionals at a particular R&D unit; and factors causing the evolution of purpose of an R&D unit.

Table 11.1 summarizes the data presented in Part I and this chapter, along with discussions with company managers. The table provides three main observations. First, in only one case of 42 R&D investments was evidence discovered that a non-R&D factor played some role in the creation of an R&D investment abroad. This particular investment involved the creation of a TTU in India by Union Carbide. However, it is important to note that the existence of "blocked funds" in India was used by the local affiliate as a supporting argument to establish the Indian TTU. They had already convinced themselves that a legitimate R&D reason existed that warranted the creation of the TTU.

In short, the preponderance of evidence indicates that non-R&D factors were not instrumental in the creation of these 42 R&D investments abroad. Other evidence supported this conclusion. For instance, CPC International's decision to create an R&D unit in Vilvoorde, Belgium, indicated that government incentives were not influential in the location of the lab. The decision to make the Vilvoorde lab the center for industrial product research for CPC Europe was implemented even though financial incentives were available from another country where similar expansion was possible. CPC managers felt that R&D operating issues were overriding. First, the Belgian location was equidistant by automobile from principal French, German, and Dutch plants, and R&D managers wanted to maximize interaction between lab and manufacturing personnel. Second, a small plant existed at the Belgian lab site, so small it could be used for pilot-plant operations. Finally, the Belgian site was located near regional headquarters and thereby would make interaction easier with European regional managers.

Second, Table 11.1 shows that non-R&D factors influenced the expansion of R&D personnel in 11 of 42 investments personnel. Yet in all 11 cases, the

TABLE 11.1

Role of Non-R&D Factors for 42 R&D Investments Made Abroad

	Non-R&D Factors Influencing the		
	Creation Decision	Expansion Decision	Evolution of Purpose Decision
Corning	None	1 unit (France)	None
Carbide*	1 of 13 units (India) blocked funds	1 unit (Canada) incentives	None
Exxon Energy	None	1 unit (France) government pressure	None
Exxon Chemical	None	None	None
IBM	None	8 units	None
CPC International	None	None	None
Otis	None	None	None

*Chemicals and plastics group only.
Source: Interviews with company managers.

presence of these non-R&D factors (public sentiment regarding the "brain drain," government agreements, pressures, and/or incentives) did not appear to be sufficient alone to cause an increase in the number of R&D professionals. Other R&D reasons also existed in all 11 cases for expansion, and it is probable that expansion would have occurred in the absence of these non-R&D factors.

One exception may have been Union Carbide's Canadian R&D unit, which apparently took advantage of Canadian incentives to run a process development program several years after the unit was created. In Corning's case, the decision to locate the European R&D unit in France at the site of Soveril's R&D unit (as opposed to some other European location) revolved partly around its desire to honor Soveril's agreement with the French government regarding the expansion of R&D activities at this location.

Non-R&D factors also may have influenced the number of R&D professionals at one R&D unit for Exxon and eight units for IBM. In Exxon's case, the French government had started suggesting that Exxon increase operations at its French unit in order to help Esso France balance its foreign payments of technology. It was not clear at the time whether Exxon was acceding to the French government's wishes.

In IBM's case, the unique market requirements of the United Kingdom, France, and Germany were responsible for the creation and early growth and evolution of the R&D units in these countries. However, their expansion during the 1950s and 1960s was also stimulated in part by IBM's concern that nationalistic outcries would occur if it helped create a "brain drain."

The growth of IBM's other five GPUs abroad were motivated possibly for the same reason, coupled with the additional requirement instituted by corporate headquarters over two decades ago to limit the growth of geographic concentrations of R&D, along with manufacturing and marketing activity (domestic or foreign) to no more than 10 percent of an area's work force population.

Finally, Table 11.1 shows that no evidence was discovered that suggests non-R&D factors caused or influenced in any way the evolution of purpose of an R&D unit abroad.

12

MANAGING R&D ABROAD

The seven organizations in this study were selected in part because of their considerable differences among each other as multinationals. Consequently, it should come as no surprise that many of the problems and opportunities they experienced with R&D abroad were also different and distinctive.

CORNING GLASS WORKS

The major problems and opportunities experienced by Corning's R&D operations abroad were associated with the transfer technology unit acquired from Soveril and subsequently the European R&D unit, created to perform indigenous technology work. The major problems were the need to reorganize the Soveril R&D unit to conform with Corning's major business-product lines; the need to honor past agreements made by Soveril with the French government regarding the Soveril R&D unit's employment level; the need to determine financial contributions of European affiliates for the European R&D unit; and the need to determine the best foreign employment composition for a unit designed to serve all European affiliates but whose present employment was composed entirely of French nationals except for the U.S. director.

The major opportunities were the opportunity of having acquired existing lab facilities with excess capacity in France that enabled expansion without any loss of time for construction of new facilities; the opportunity of acquiring a foreign R&D unit in one business where Corning U.S. did not have direct technology, marketing, or production experience (for example, sun glasses); the opportunity of acquiring R&D personnel with specialized technology skills (in optical melting) not possessed by Corning's other R&D units; and the chance to obtain some corporate funds and R&D domestic management resources for the European unit to aid its expansion.

UNION CARBIDE

Several major problems were experienced by Union Carbide's R&D units abroad. First, nonexistent or weak spatial associations with production and/or marketing personnel and customers were a source of difficulty for some TTUs. For instance, the Hythe U.K. unit was ultimately disbanded because its location was not favorable vis-a-vis regional product managers located in Geneva and chemical customers on the Continent. The Swiss unit's spatial separation from production operations was also a drawback but was apparently surmountable by tying the unit administratively with the Antwerp process unit. The neutrality of the unit's location vis-a-vis EEC and EFTA joint-venture subsidiaries did not stimulate these groups to utilize the Swiss unit's facilities.

Second, the Swiss TTU experienced some significant problems as it increased its size and foreign employee mix. The transfer of personnel from the Hythe lab to the Swiss unit, along with the unit's own growth, contributed to the Swiss unit's need to expand its facilities. This was a time-consuming proposition due to the particular problems of building in Switzerland. Also, the transfer of British personnel from the Hythe lab increased the foreign composition of employees at the Swiss unit. This increased existing problems of compensation (based on different national expectations, especially regarding housing) along with some cultural communications difficulties.

Third, the relatively small size of Union Carbide's European chemicals and plastics market (with much of it based on U.S. exports) made it difficult (impossible in some cases) for the R&D manager of the Swiss TTU to plan long-term R&D projects. European business managers were reluctant to fund R&D projects for U.S. exports because their long-term availability was dependent on U.S. consumption. If a large U.S. customer (like RCA) suddenly increased its purchases of a product from Union Carbide, exports to Europe could be reduced or stopped suddenly. Consequently, R&D managers felt the unit's growth was dependent to some extent on Union Carbide's growth in European production.

Fourth, the director of the Swiss TTU felt the unit's present small size kept it out of certain technology areas where it faced stiff competition from other European-based multinationals with much larger R&D operations in those areas. And fifth, the relatively recent creation and acquisition of several foreign R&D units led senior Union Carbide R&D and business managers in the United States and abroad to spend time considering the coordination of domestic and foreign R&D operations.

However, several advantages or opportunities were also experienced by Union Carbide's foreign TTUs. First, some units had the opportunity to develop close manufacturing and/or marketing associations forced by technical service work that familiarized their R&D professionals with Union Carbide's businesses. Also, some units had the opportunity to play important roles in the transfer of technology from the domestic parent. For instance, the Swiss unit worked on projects whose products accounted for 25 percent of Union Carbide Europe's total chemicals and plastics sales in 1973.

The Swiss unit also had the opportunity to work on increasingly more sophisticated projects to develop new and improved products for special European markets whose output would be used in several European nations. The same unit also had the opportunity to work on projects to develop new and improved products primarily for the European region but that might be useful to the domestic parent (for example, a new brake fluid had been transferred to the United States from the Swiss R&D unit).

Union Carbide's Canadian chemicals unit had the opportunity to participate in an international program where several domestic and foreign units worked on portions of a production scale-up for an experimental process. Also, some Union Carbide R&D managers felt that the New Delhi unit had the opportunity to expand quickly in terms of transfer technology work into indigenous technology work because R&D competition was weak in India. Additionally, the company had growing manufacturing interests in agricultural chemicals that could support the expansion.

Last, the addition of the three BXL units had quickly augmented Union Carbide's R&D resources in plastics which were capable of meeting the particular product needs of the European business.

Several problems and opportunities were also mentioned regarding Union Carbide's experience with its Belgian CTU. The Belgian unit was isolated spatially from the European business, and U.S. managers sent abroad could not orient the activities of the unit's R&D professionals to produce relevant results for the company. According to Union Carbide officials, the Belgian unit had no impact on international technology flows, despite a dozen years of exploratory research activity. The unit never produced any output utilized by the parent or its foreign affiliates. And the legal settlement of foreign employee contracts of the Belgian unit was difficult, expensive, and time consuming because of Belgian law.

The principal opportunity noted by interviewees regarding the Belgian unit was the hiring of several top European scientists, a few of whom were retained by the company and relocated to the United States after the Belgian R&D unit was disbanded.

EXXON CORPORATION

The main problems experienced by Exxon's foreign R&D units were some difficulties during the 1960s in working out financing arrangements with the corporate parent and Exxon U.S. regarding projects of mutual interest; obtaining funds when a seller's market existed in Europe and marketing did not need R&D's help to move its products; and implementing areas of specialization among the four European R&D units.

The principal opportunities experienced by Exxon's foreign R&D units were their growing ability to perform R&D expressly for the parent and their regional businesses as they expanded in size and particularly since they adopted product-business areas of technical specialization; and their chance to assign larger numbers of R&D professionals to perform R&D expressly for the parent or their regional businesses since the energy crisis, which reduced the number of R&D resources needed for technical service activities.

EXXON CHEMICAL COMPANY

Before the creation of the Belgian unit, R&D projects were assigned to units whose locations had been determined chiefly by the location of preexisting petroleum R&D facilities and talent. However, the logic underlying the location of the petroleum R&D units did not apply across the board to chemicals R&D activity. Senior managers at Exxon Chemical felt they needed tighter coordination among R&D professionals and between them and regional business managers.

The creation and evolution of the Belgian unit began solving this problem; however, other difficulties arose because the unit was composed of foreign nationals from seven different countries. There were problems with different patterns of compensation, which varied by national affiliate within the company; problems matching administrative needs with language capabilities of R&D professionals; and problems involving culture gaps between personnel. Besides these difficulties, all the R&D units encountered problems when trying to transfer new or improved technology developed by them. Written reports were found to be insufficient. Instead, they began to travel more to promote transfers themselves; and to form technology transfer teams for important projects, composed of R&D, engineering, manufacturing, and marketing people.

The major opportunities were the special advantages of having an older, $26 billion sister business in petroleum with preexisting foreign R&D facilities and personnel that allowed foreign R&D operations earlier than otherwise would have been possible; a new business organization that gave eight product managers worldwide responsibility for technology development and increased the possibility that the foreign units could receive funding for global application; and the disbandment of the German unit, which was neither difficult nor painful because of the association with the petroleum R&D unit (chemicals people were either reassigned to oil R&D projects or transferred to the Belgian unit if they elected to move).

IBM

Two sets of problems and opportunitis existed for IBM: those relating to IBM's eight GPUs that were part of the World Trade Corporation and those relating to the CTU in Switzerland that was part of the research division.

The main problems noted by IBM executives regarding its GPUs related to difficulties experienced by the U.K., German, and French units. These problems were getting research support before the Swiss unit was created because U.S. support was too hard to obtain because travel time was much longer during the 1950s; controlling new products developed by the early foreign units who wanted their own computer lines and models; controlling domestic units from infringing on product-technology areas assigned by corporate headquarters to the foreign units when this practice began during the early 1960s; and the need to develop an independent corporate position mainly to develop long-term R&D policy but also to monitor proliferating R&D operations in the United States and abroad.

The main opportunities realized by the eight GPUs were the chance to obtain outstanding R&D talent, some of whom would not have left their home countries to work in the United States or other countries; the opportunity to develop a strong sense of community at R&D units composed entirely of local nationals; the opportunity to concentrate their resources and build skills in specific product-technology areas; the opportunity to develop R&D resources that were ready and able to generate special new and improved products for their foreign markets when corporate policy deemed necessary; and the opportunity to play a critical role in their generation of computer technology.

The principal problems experienced by the corporate Swiss unit were stimulating and motivating top scientists and engineers in general directions desired by IBM; in short, managing them since corporate research managers did not believe in the unstructured, hands-off policy toward creative, scientific personnel; and the fact that the generation and flow of ideas required great technical vitality and a continual need for new people and regeneration programs for the older people.

The main opportunities included the chance to perform exploratory research under a corporate charter from the unit's creation that left no doubt that their activities were long term in nature and corporate managers had no expectations that the unit would have to develop a new kind of computer for immediate commercialization; and the chance to work on specific projects that were well defined. A third opportunity was the chance to work in long-term research areas that were not isolated from European development activities nor from domestic corporate research. That is, the Swiss unit was tied to corporate R&D in the United States through a program of shared projects that the director of the Swiss unit felt lowered the critical mass factor facing the unit. R&D managers at IBM felt these three opportunities were chiefly responsible for the Swiss unit's ability to contribute new technology in solid state electronics and physics for word and data processing application.

CPC INTERNATIONAL

The main problems noted by CPC managers regarding R&D abroad were that the disbanded U.K. TTU was not located close enough to the center of European business in industrial products in order to justify further expansion of the unit; that the Japanese CTU was not bonded spatially with manufacturing operations, making association of the unit's activities with the Far East affiliate's operations more difficult; and that the growth of R&D abroad required expanding the responsibility of the corporate vice president of R&D for coordinating R&D operations worldwide.

CPC managers noted three princiapl opportunities regarding R&D activities abroad: the opportunity to gain experience transferring technology developed by the Belgian, German (Heilbronn), and Swiss units to other European units; the existence of an R&D capability in consumer business at the German (Heilbronn) and Swiss units that is considered stronger than CPC's domestic capability; and the opportunity to utilize technology developed mainly by the Belgian, German (Heilbronn), and Swiss units throughout the world, as these units have accounted for most of 25 major products and processes developed outside the United States along with over 100 patents developed abroad that have been utilized worldwide.

OTIS ELEVATOR

No problems or opportunities were noted by Otis Elevator managers regarding its acquired R&D units. The principal problems and opportunities experienced by Otis Elevator's R&D operations abroad were associated with the evolution of five TTUs created by Otis Elevator. An early problem for four of these five R&D units was preventing innovative-minded engineers from continually redesigning products and components developed by the U.S. parent. The addition of new product development activities by these units provided an opportunity to channel these innovative minds in a direction more congruent with company goals. As this work expanded, European-level managers saw the opportunity to have each unit specialize in a particular technological area. This was implemented, though not without considerable difficulty, in transferring people across European borders in order to concentrate similar technological skills and in getting units to give up technological capabilities to other units.

Also, joint project work revealed that the four TTUs in Europe lacked a common technical language. A considerable amount of R&D work was performed during the 1960s but all reports had been written in the national language of R&D unit. All reports were subsequently written in two languages, one of which was English.

Considerable problems were encountered in implementating engineering and R&D specialization because each engineering department had done everything in the past. For instance, Otis Europe was faced with the need to cut back on electronics specialists in France if it wanted to do all electronics work in Germany. Otis Europe experimented with moving some specialists across national borders in Europe but such permanent transfers did not work out. People stayed for 12 to 18 months and then returned home.

13

R&D ABROAD AND
TECHNOLOGY TRANSFERS

The process of R&D abroad for U.S. multinationals appears to start with the establishment of transfer technology units. Consequently, one would expect that the dominant international pattern of technology transfer would be from the U.S. parent to foreign subsidiaries as long as transfer technology work was the dominant form of R&D aboard.

Table 13.1 summarizes the impact of R&D abroad for each multinational. It reinforces the expectation that the U.S. parent will supply nearly all new product-process technology as long as transfer technology work is the dominant form of R&D abroad. However, the table also shows that this initial pattern of technology transfer changes once indigenous technology and/or global product work begin abroad.

Overall, Table 13.1 indicates, along with observations for each multinational presented in the remainder of this chapter, four main points about the impact of R&D abroad on international patterns of technology transfer:

1. Significant new product-process work has been achieved by R&D abroad in five of the seven multinationals in this study.
2. International transfers of this product and process technology have occurred, mostly within the European region, for five of the seven multinationals once ITUs or GPUs were established abroad.
3. Third-world subsidiaries of the same five multinationals have been obtaining growing amounts of new products and processes from European subsidiaries once ITUs or GPUs were established abroad.

TABLE 13.1

Impact of R&D Abroad on International Patterns of Technology Transfers

U.S. Multinational	Impact	Other Observations
Corning Glass Works	Negligible. Some transfers expected between European subsidiaries in near future. Nearly all technology still supplied by U.S. parent.	Most recent involvement with R&D abroad of seven multinationals.
Union Carbide (chemicals and plastics only)	Negligible. U.S. parent still supplies nearly all technology.	Limited R&D resources abroad. Still all TTUs but changing to ITUs.
Exxon Corporation (energy)	Initial pattern changing. ITUs abroad have produced new products and processes for regional and inter-regional transfer to third world subsidiaries.	Most R&D work is oriented toward indigenous markets. A few global-product-process products recently under-way.
Exxon Chemical Company	Initial pattern changing. ITUs are producing new pro-ducts and processes for intra-European transfer as well.	
IBM	Significant flows of product technology worldwide, in-cluding to the United States once GPUs were created.	Appears to be movement toward more indigenous technology work, which may suggest drop in transfers to U.S. in the future.
CPC International	Initial pattern changed. European ITUs supply much of own technology needs plus needs of third-world nations.	Some flows to United States, although domestic labs and acquisitions still supply most new technology for U.S. businesses.
Otis Elevator	Initial pattern changed. European ITUs supply most of their technology needs plus needs of third-world nations.	Very little transfers to the United States.

4. Some significant transfers of products and processes have come to the United States from global product units; however, in the absence of GPUs, technology transfers to the United States have been minimal in the view of interviewees.

CORNING GLASS WORKS

According to Corning managers, foreign R&D activities had not yet influenced the existing international pattern of technology transfers. Under this pattern, the domestic parent supplied practically all international transfers of technology. However, the new European R&D unit was expected to alter intra-European transfers within the near future.

The approximate number of R&D projects handled by the European lab (formerly the Soveril lab) at Avon since 1970 are listed in Table 13.2 along with

TABLE 13.2

Corning Glass Works: Number of R&D Projects Performed by the Avon Lab, 1970-74, and the Prospective Users of Technology by Geographic Location

Users	Number of Projects				
	1970	1971	1972	1973	1974
National only	20	17	15	11	9
European	0	0	0	1	1
United States	0	0	0	0	0
Other	0	0	0	0	0
Total	20	17	15	12	10

Source: Company interviews.

the prospective users of the technology whether or not the project was completed or successful. Some of these projects are the same across the five years shown, some are new, and some have been terminated or completed. The total number of projects has decreased because Corning believed the average project size was too small given the lab's resources. In every case except one, the actual or intended user of an approved project was Soveril. The lone exception is a 1973 project that will be used, if successful, by most of the European subsidiaries.

The expectation is that more such European-oriented projects will be initiated in the near future.

Equally significant was the shift into some exploratory research (leading into existing business) and a small amount of new high-risk business R&D performed by the Avon lab. In the past, R&D performed at Avon was entirely existing business R&D, generally minor product modifications work in optics. The lab's funds were devoted primarily to product-related work (90 percent) as opposed to process work (10 percent). With the possible exception of some evolutionary work in glass melting over the last six years, the lab's personnel claimed no major product or process breakthroughs. Its funds, until recently, were not geared toward the development of major innovations that might have international technology transfer potential. R&D was not stressed under Soveril except for minor product modification work and most of the lab's professionals saw the new association with Corning as an opportunity to perform more serious, higher impact R&D work.

UNION CARBIDE

Union Carbide managers felt that foreign R&D operations had not altered the company's international pattern of technology flow whereby the U.S. parent supplied nearly all technology utilized abroad. However, the growth of indigenous R&D operations at the Swiss unit suggested that Union Carbide Europe might alter this dominant pattern slightly by supplying some of its own technology needs in the near future.

EXXON CORPORATION

In the past, Exxon's major international technology flows emanated almost entirely from the United States and were transferred abroad directly to the technology's end user. Exxon had obtained some important process technology abroad through licensing agreements, but R&D-sponsored technology generated through internal R&D was developed almost exclusively by domestic R&D units and then transferred to foreign affiliates. In fact, this direction of technology flows was generated because whenever an important discovery was made abroad, the project was transferred to the United States. For instance, a major transfer of new product technology came to the United States from R&D work done in Canada about a dozen years ago. With the discovery of Paradyne 20, a whole new line of fuels was developed. At first, the project had a specific Canadian orientation but when the ramifications of the initial breakthroughs were comprehended, the project and the discovering scientist were transferred to Linden, N.J., where greater R&D resources were available.

This basic or exclusively U.S.-led pattern of international technology transfer appeared to be changing by 1974, although the U.S. R&D units still remained the primary source of new major international technology flows within the Exxon system. These units possessed the overwhelming share of resources for long-term exploratory research that were being devoted to the generation of new product-process technology.

> The Baytown, Texas, Laboratories, work proceeds toward an economical process for converting coal into liquid and gaseous fuels. At the Baton Route, Louisiana, Laboratories, a process is being evaluated to refine synthetic crude from Canadian tar sands. Participation is underway also in an industry-academic program to explore the concept of laser-induced nuclear fusion.
>
>   Exploratory, fundamental and development studies continue on a broad front that goes beyond the classical business boundaries. Interdisciplinary approaches are encouraged to develop new technology for ongoing corporate interests and for new, long-term, high-risk projects such as enzyme catalysis, solar energy, and bioconversions.[1]

Nevertheless, the basic pattern was changing. For instance, the English lab was working on an important R&D project (clean-burning oil) because it had the technology capability underlying the project in fluid-bed operations. The project results would be transferred to the United States.

In Canada, Exxon's affiliate, Imperial Oil, was responsible for developing new process technology to produce and refine heavy oil reserves. Lab and pilot-plant operations encouraged Imperil to build a 4,000 barrel-per-day plant in a scale-up application of the new technology. If successful, the technology would be transferred to other nations that, like Canada, have large heavy oil reserves that are not economical to produce under existing recovery techniques.

A special review of Mont San Aignan's international technology flows substantiated this new transfer pattern, which saw the European labs primarily supplying much of their own technology, supplying the U.S. parent with a small amount of new technology, and supplying third-world country affiliates with technology that previously could only have been available from U.S. R&D units. For example, a new process device was being developed by Mont San Aignan to make refinery furnaces more efficient fuel burners. The new device would be used worldwide by Exxon (including the United States) and was expected to yield considerable savings in energy consumption. Also, another burner was being developed that would burn pollution free, which, if successful, would have worldwide transfer application.

In 1974, the Mont San Aignan lab was working on 88 R&D projects. Nineteen projects were mutualized petroleum R&D projects. Twenty-five were mutualized chemicals R&D projects. The remaining 44 projects were "self"-R&D

petroleum projects for Esso France. The expectation for international transfer applications, based on past experience, was that 10 percent of these projects would probably be picked up by other affiliates, especially in Scandanavia, North Africa, South Africa, and Japan. The primary users of these mutualized petroleum projects would be European affiliates, although one or two of these would be transferred to the United States and Canada.

Other examples of important cross-national technology transfers from the French lab were the development of special asphalt formulations utilized in a European road construction computer program, a special sulfanate (synthetic oil) process, a new white oil process (without using sulfuric acid), and an improved process for insulating oils for higher quality.

The new alternatives for obtaining technology by foreign affiliates without their own R&D capability was another indication of change in Exxon's original pattern of international technology flows. For instance, Exxon Brazil had no technical capability and at one time had no choice but to go to the relevant U.S. lab to obtain needed technology. Technology transfers, in short, no longer moved entirely from the United States direct to a foreign technology user. The European labs were now supplying modified U.S. technology or their own produced technology in a few instances. Now Brazil went through its Latin American regional headquarters where marketing technical services recommended which lab Brazil should approach in the United States, Canada, or Europe.

EXXON CHEMICAL COMPANY

The emergence of the Bruke-Hartke legislation in the early 1970s caused Exxon Chemical to examine the significance of foreign technology and its association with it. Top R&D managers feel they did not have an adequate understanding of the nature and direction of technology transfers between U.S. R&D units and R&D units in other nations.

In the past, nearly all R&D-sponsored international technology transfers flowed from the United States to technology users abroad. Little technology flowed back to the United States. Some Paramins technology was developed in Canada about a dozen years ago and the English and French R&D units can point to one or two projects that eventually were transferred to the United States.

The foreign R&D units created during the 1950s were established primarily to furnish local plant or marketing technical assistance for the national affiliates and their customers. However, the European R&D units began doing cross-national, regional R&D work for affiliates in other European countries during the early 1960s. Again, this was mostly technical assistance work. Much of this technical work for other European affiliates was consolidated at the Belgium lab after it was created in 1965. Since then, the Belgium lab has steadily increased the number of R&D projects it performs for cross-national transfer to non-Belgian affiliates.

Today, the Belgium lab has approximately 30 current R&D projects (not including technical service projects). The majority of R&D projects are for European application. A few projects will be transferred and used in the United States and Japan. The work includes both product and process R&D projects. The general expectation is that the amount of R&D-sponsored technology transfers used outside Europe will increase significantly as new manufacturing patterns emerge in Europe. Exxon managers feel these new patterns will develop because Europe as a whole is almost the equivalent of the U.S. market in chemicals.

The general experience of the Belgium lab regarding the actual transfer of technology is that it stopped trying to implement transfers through written reports. It found that fruitful technology transfers required the physical transfer of personnel to the United States, Japan, or wherever the technology was to be applied. It also found it meant having transfer teams composed of R&D, engineering, manufacturing, and marketing representatives.

The point was made, however, that successful R&D technology transfers were a function of more than any single R&D unit's effort. The organization of the entire system was also a critical factor. For instance, Exxon Chemical's 1972 structural reorganization along product technology lines seemed to enhance the possibility for cross-discipline teamwork and communication for both technology generation and transfer. Before the reorganization, an in-house study determined that the level of communication was very poor regarding technology generation and technology transfer.

> "We confirmed our suspicion that barriers existed," says Holmes [the study's director]. Research and engineering were not communicating sufficiently. The plant people were not communicating with either on a regular basis. So the new grouping was structured. "We are putting together all the process people from research and from engineering. We're making research more engineering-oriented and engineering more research-oriented."[2]

IBM

Today few firms spend as much money on R&D as IBM ($730 million in 1973). The corporation believes these large expenditures are paying off, so much so that, in 1974, Gilbert Jones, IBM's vice chairman, could say that the corporation's competitive abilities were no longer based primarily on its marketing prowess but also on its ability to generate new technology. Jones further noted, "What most people do not recognize, however, is that critical contributions in the generation of technology have occurred abroad in IBM's foreign labs."[3]

Table 13.3 presents some of the major technology transfers of IBM's European development labs to the United States. The list focuses only on major contributions occurring over roughly the last half dozen years, that is, only those transfers having corporate impact as opposed to intranational and/or intra-European transfers of technology.

These technology transfers are impressive for two reasons. First, they understate the actual volume of international technology flows from R&D abroad to the U.S. parent. Important transfers from the Zurich research lab are excluded as are transfers from R&D work on the System/360 and earlier work. Second, these technology flows have been realized over a relatively short time span, since IBM began investing earnestly in R&D abroad less than 20 years ago. In fact, IBM has not always been interested, much less heavily committed to R&D either in the United States or abroad. Large R&D allocations began only two decades ago during the early 1950s when IBM's decision to enter the computer market triggered an important shift in its R&D tradition and its R&D organization.

In addition to international transfers made by the GPUs, technology was also transferred by the Zurich corporate technology unit, primarily to the United States. Over the 1970-73 period, these transfers included none to the national (Swiss) subsidiary, none to non-European subsidiaries, one major transfer to the French and German companies for components R&D work in voice communications, and several major transfers to the United States of new technology and devices in solid state electronics and materials storage.

CPC INTERNATIONAL

Over CPC's entire history, the lion's share of major technology transfers across national borders has emanated primarily from the United States and flowed to Europe and the rest of the world. This initial pattern of international technology flows seems to be undergoing fundamental changes over the last half dozen years. The new pattern sees Europe playing a major role in supplying R&D-sponsored technology to other European nations, Latin America, and the Far East, and an important, though not dominant, role in providing the parent's domestic businesses with both product and process technology.

R&D-sponsored technology generated in Europe resulted in transfers mainly between European countries and from Europe to other foreign regions for several reasons. One reason was simply that R&D-sponsored technology developed in Europe was designed primarily for the European business. For instance, no projects were being performed in Europe expressly for the United States. (However, three ongoing European industrial R&D projects were being reviewed by U.S. managers in 1974 for potential financial backing.)

TABLE 13.3

IBM: International Technology Transfers from IBM's Global Product Units Abroad

Boeblingen, German lab
  Design of small computer systems—developed the System/370 Model 115 and Model 125
  Semiconductors; developed device for better transistor packing on silicon chips—used in System/370, Models 115, 125, 158, and 168
  System software
  High-speed printers

Hursley, English lab
  Intermediate computer systems—developed most of system 370/Model 135
  Specialized terminals—developed for banking and airline industries
  Disk storage—developed devices for small computer application and terminal subsystems
  Software—major role in developing PL/I language; also helped develop an advanced control program for operating large networks of terminals linked to central computer

La Gaude, French lab
  Programming—conversion aids of help users move across different computer systems
  Note: The French lab is also doing original R&D work in telecommunications. Though the output of this work has not been transferred or utilized in the United States, it could be critical to the industry and IBM, especially if IBM enters the U.S. telecommunications field. Basic technology, programming, and system design work are being performed by the French lab for communications-based systems

Lidingo, Swedish lab
  Software—developed sort programs and program products used in data base systems
  Hardware—developed communications equipment (note: used mainly in Europe)

Uithoorn, Dutch lab
  Software—disk operating system, a control program for small and medium-sized System/370 computers
  Hardware—developed first optical reading technology and optical reader sorters

Vienna, Austrian lab
  Software—performed theoretical and definitional work for computer languages, on data processing principals, and language compiler structures. Helped to develop PL/I

117

Also, CPC's European executives felt the adoption of a spearhead approach to technology transfer was a major reason for the low resistance encountered in transferring technology developed in Europe on an intra-European basis or to other foreign regions. The spearhead concept involved designating and building up one country affiliate in terms of technical capability and receptivity to innovation. New ideas and methods were introduced in this country first and adoption in other countries was accelerated and facilitated via the demonstration effect.

A third reason was that CPC Europe assumed a major role in technology transfers to other foreign regions because CPC Europe wanted to maximize investment returns on its indigenously developed technology by tying other regions to its product-process technology as much as possible; and the market-technology gap in terms of market needs versus technology matchup was smaller apparently between Europe and some other foreign countries than between these latter countries and the United States.

The flow of technology moving from Europe to the United States has been important, although the domestic businesses remained their own primary source of new product-process technology. However, CPC Europe believed that it had an existing potential to provide CPC's domestic businesses with more technology in the future. With each passing year, CPC's technology is assuming a stronger multinational flavor. Thus far

> More than 100 patents developed by affiliates in Belgium, Britain, Germany, Japan and elsewhere have become available for its U.S. operations. More than 25 major process and technological improvements in the company's worldwide operations were developed by its affiliates outside the United States.[4]

OTIS ELEVATOR

Since 1971 (and this date is only a rough cutoff point), Otis's international trade and foreign expansion has been based on modified U.S. technology and new European product technology produced by its European-based R&D and engineering units. This change began in 1960 when some European managers started questioning Otis's strategy of being the sole custom-made, high-rise elevator-escalator producer for Europe and other lower income markets. Instead they felt Otis should attain scale economies and mass produce small standardized elevators in head-to-head competition with local producers for the apartment and small building market.

All existing business R&D activities performed by Otis's European subsidiaries over the last 15 years rested on fundamental technology transferred abroad by Otis U.S. or other U.S. firms outside the elevator-escalator business (for example, RCA in solid state electronics). However, the existence of evolving R&D operations in Europe had spawned some important changes in international technology flows within Europe and the rest of the international division.

Since Otis's earliest days of foreign involvement, technology had always been transferred abroad without any modification. However, when the European and Japanese markets for elevator-escalators became substantially different from Otis's U.S. market during the late 1950s, unofficial redesign work started being performed in Europe. During the 1960s, product modification work was permitted by the parent. This work evolved from relatively minor product change to major production modifications, including the development of a unique European product line.

By 1974, Otis Europe was providing most of the product-process technology for Europe and the rest of the international division. Approximately 64 active "European" R&D projects were being performed by four European ITUs that could have a potential technology impact for more than a single national subsidiary. These same four European units had transferred most of the technology within the international division (that is, to other European and non-European nations except the United States and Canada) over the 1965-74 period.

Only one instance existed of technology being transferred back to the United States; however, the feeling was that these flows could increase substantially in the near future. The development of small elevator technology in Europe offered the parent the opportunity to apply this technology to a growing U.S. market that it had not considered attractive in the past.

Specifically, R&D activities in Europe over the last half dozen years resulted in the development of product modifications in escalator technology in Germany that were transferred all over the world, including the United States; the development of a European line of small elevators for office buildings and hospitals in 1971 (Series 80) that have been exported to the rest of the world, except the United States (however, Otis U.S. might use some of this technology in the future); the current development of several other European R&D projects that the United States might use in the future but that would be used initially in Europe and the rest of the world; and the development of a line of small elevators for apartment buildings in 1973 for Europe, other foreign regions, and possibly the United States. (The smallest elevator Otis U.S. made carried 12 passengers versus Europe's 4-passenger model.)

Overall, Otis executives in the United States and abroad felt the company's European R&D work was different and supplementary to U.S. R&D work. The general consensus was that Otis would be considerably weaker as a multinational enterprise without the benefit of its European R&D activities because Otis Europe now supplies the technology that has kept the international part of the business competitive, and this represented more than half the total system sales and earnings.

NOTES

1. Exxon Corporation, *Exxon* (New York: inhouse publication), p. 12.

2. "New Move Puts Technology on the Product Line," *Chemical Week* (August 15, 1973): p. 29.

3. Jones, Gilbert, observations made during speech at Harvard Business School, Cambridge, Mass., spring, 1974.

4. CPC International, Inc., *Almanac 1973-74: A Directory and Fact Book of A Multinational Food Company* (Englewood Cliffs, N.J., 1974), p. 9.

CONCLUSIONS:
ADMINISTRATIVE IMPLICATIONS

Although the present study is exploratory and limited in sample size, an analysis of the 47 R&D units created outside the United States by seven U.S.-based multinational organizations provides evidence for several hypotheses about the process of R&D abroad. Some tentative conclusions can be drawn that can serve as a basis for future hypothesis testing and as a basis for administrative policy, at least until a more comprehensive analysis is available.

THE R&D PROCESS ABROAD

All four types of foreign R&D units—transfer technology units, indigenous technology units, global product units, and corporate technology units—were established after the U.S.-based parent organization made the decision to create new R&D activities abroad. Yet, almost one quarter of the R&D units in the sample were obtained by the parent organizations when they acquired foreign companies. In all instances, these acquisitions were for non-R&D related reasons. Acquired R&D units included indigenous technology units and transfer technology units (in the latter case, because the U.S. parent licensed technology to the foreign company before acquiring it). Since these acquired R&D units were not operating as a result of specific corporate decisions to commence R&D operations abroad, they are not included in the ensuing analysis.

The 42 R&D units created abroad fell into two categories. The first was related to foreign manufacturing investment; the second to the scientific and technological strategy of the U.S. corporate parent. R&D units in the first category (transfer technology units, indigenous technology units, and global product units) were dependent on foreign manufacturing investments for their creation, growth, and evolution of R&D purpose. They were often located within or near foreign operating units, particularly manufacturing operations. Together, they accounted for more than 90 percent of the foreign R&D investments in this study. R&D units in the second category (corporate technology units) represented less than 10 percent of the foreign R&D investments in this study. Corporate technology units initially had no connection with the foreign investment process, and were administratively and geographically isolated from other foreign operations. However, the CTUs analyzed in this study became associated with foreign investment activity when they changed their R&D purpose.

The majority of R&D investments associated with the first category (82 percent) were created by managers in foreign manufacturing subsidiaries to aid in the transfer of technology provided by the U.S. parent. The creation of TTUs was essentially a decentralized phenomenon, involving a few R&D professionals who performed slight product-process alterations for market and plant customers within a foreign national market.

TTUs were created when product-process technology was unstandardized. They were concentrated in countries that were the sites of the U.S. parent's earliest manufacturing investments, where technology tended to be most unsettled. Most of these units were in European nations or Canada, where the parent companies' earliest manufacturing operations were established. In other words, transfer technology units were the first form of foreign R&D investment. Directors of TTUs reported directly to functional managers outside the R&D area and performed R&D activity entirely in support of existing business.

However, the transfer technology units in this study followed a definite trend. They often evolved into indigenous technology units, because they could not keep their best R&D professionals unless they could offer the opportunity for more challenging R&D projects. Whether or not an R&D director was able to provide such opportunities depended on the need of foreign operating managers for new and/or improved products. Indigenous technology projects were commissioned at transfer technology units when foreign managers perceived that the new and improved products transferred from the United States were insufficient to sustain the growth of the foreign subsidiary. Again, the data suggest this was more likely to occur in the older, larger European subsidiaries.

Several characteristics of transfer technology units changed when they became indigenous technology units. First, the directors of the units began to report to general managers, often at the regional headquarters level. Second, the units' responsibility grew beyond the national level to the regional level. Third, in some cases, the strategic composition of R&D expenditures included R&D to develop new high-risk business and/or some exploratory research for existing or new high-risk business.

At this point in the evolutionary process, foreign R&D operations were still usually controlled by general managers located abroad. However, corporate managers in the United States began to intervene either to take advantage of technological skills developed at R&D units abroad (without changing the R&D unit's primary purpose) or to redirect the activities of foreign R&D units away from indigenous technology work and toward the development of new products for use in the United States and foreign markets.

As a result of the latter type of intervention, three indigenous technology units evolved into global product units. These three units began to perform global product work when the capacity of existing R&D units within the parent organization's domestic operations was insufficient to develop new products and could not be expanded. These global product units were located at major manufacturing-marketing centers in order to promote communications among production, engineering, sales, and R&D personnel. The R&D directors of these units reported to both national subsidiary and corporate managers. Their geographic level of responsibility was global and the strategic composition of R&D expenditures was entirely in support of existing business.

The increase in the number of R&D professionals at foreign R&D units usually occurred when the units changed their R&D purpose, regardless of their original purpose, age, or location. For example, transfer technology units added, on the average, one R&D professional per year. Increases were small because one R&D professional could handle several technical service projects. The unit also became more proficient over time. However, the number of R&D professionals increased considerably when a transfer technology unit switched to indigenous technology work. This was because new and improved product-process projects were more complex and more expensive than technical service activities. Once the transition into indigenous technology work was complete, increases in R&D professional staff declined. Nevertheless, the annual rate of increase was usually twice as large as the transfer technology units (that is, two R&D professionals per year).

The evidence, though limited, suggests that the size of an R&D unit will be especially large if it evolves into (or is created as) a global product unit. The development of new products for simultaneous U.S. and foreign production offers the prospect of extremely large market size, sales, and earnings. Potentially high returns, coupled with the need for rapid development of new products (possibly concurrent with domestic development), pushes project development costs up, as management plans greater specialization for R&D tasks to ensure success.

The creation of corporate technology units was unrelated to foreign investment in manufacturing. Their communication channels and the strategic types of R&D they performed differed from those of other R&D units. For example, CTU directors reported only to corporate officials, and their units usually performed exploratory research in support of existing businesses or to develop new high-risk business.

The decision leading to the creation of the corporate technology units in this study grew out of top corporate executives' concern that their U.S.-based

R&D units were not generating the long-range technology their companies would need in the future to remain competitive. Concurrently, related technological advances made by foreign scientists caught the attention of top corporate executives abroad. When companies sought to hire foreign-based scientists to work in R&D units in the United States, they were often turned down. Scientists refused to move to the United States. Corporate executives then decided to create corporate technology units abroad in or near the scientists' home countries.

Evidence suggests that corporate technology units were unproductive when they did not focus a sufficient number of R&D resources within particular technology-business areas that were related to the parents' existing business operations. The end result was small, or no, increases in the numbers of R&D professionals and a high probability that the corporate technology unit would be dissolved unless it changed its purpose.

In order to change its purpose, the corporate technology unit could become associated with the R&D needs of foreign investment operations. This was difficult to achieve in practice because the location of corporate technology units was separate spatially and often isolated from the organization's foreign manufacturing operations. Two corporate technology units in this study have become involved in indigenous technology work for foreign affiliates to perform R&D in support of existing businesses or to provide exploratory research capability.

Overall, the R&D units in this study exhibited movement toward the performance of indigenous technology work in the absence, or relinquishing, of corporate control, regardless of an R&D unit's original purpose. This tendency to evolve into indigenous technology work had an important impact on subsequent international patterns of technology transfer. Individual case histories of each organization showed that changes in patterns of international technology transfers occurred as indigenous technology units evolved abroad. Since these units existed mainly in Europe, European subsidiaries became less dependent on the U.S. parent for new technology. They also began to supply other foreign subsidiaries with new product-process technology. In a few instances, there were reverse flows of technology to the United States, but these seemed to be important only when global product units were created abroad or when they evolved from other types of R&D units.

ADMINISTRATIVE IMPLICATIONS OF THE R&D PROCESS ABROAD

The R&D process for R&D units located outside the United States suggests a number of administrative guidelines for multinational managers. One way to explain these guidelines is to explain the change in attitudes of general managers about the importance of the foreign R&D investments in this study.

At first, this attitude was one of almost total disregard (that is, when transfer technology units were being created), since the creation of TTUs had very little impact on operations outside the manufacturing subsidiaries where they were established. Yet transfer technology units in this study exhibited an unmistakable tendency to move into considerably more sophisticated R&D activities, projects that had substantial impact on future multinational operations. When the size of foreign R&D operations became larger, the result was a complete shift in general management's attitude, from disinterest to careful scrutiny. This scrutiny revealed that there were some R&D units that were not organized to bring the most benefit to the system.

These few observations suggest that an organization's initial administrative posture toward R&D abroad should change: future possibilities should be considered before they occur. Once this strategic posture is adopted, general managers at corporate and operating levels must reach some consensus on whether or not they wish to influence the evolutionary direction, and hence the size and nature, of R&D operations abroad. Otherwise, in the absence of a managerial decision, the evolution of R&D abroad will take a specific direction, toward the development of new and improved products and processes expressly for the foreign market.

As soon as foreign manufacturing operations begin, planners should note whether products and processes are highly standardized or whether they are subject to small, customizing alterations. Where the latter condition exists, a need may exist for transfer technology units. At this point, R&D planners should analyze the system's needs for new and improved product-process innovation and how these needs relate to the location of R&D activity.

This kind of early analysis is desirable because evidence suggests that the changeover to indigenous technology work commits major foreign subsidiaries to businesses that, over time, become increasingly different from the domestic businesses of the parent organization. The evolutionary process also results in the flow of technology to other foreign subsidiaries in smaller market nations. This may, or may not, be the goal of the corporate parent.

The decision to speed, retard, or alter this natural evolutionary movement of R&D abroad should depend on how an organization can best use its R&D resources as a competitive weapon. Findings suggest that foreign R&D projects can be considered in terms of potential national, regional, and/or world product-markets for particular kinds of product-process development.

One way to analyze these alternatives is to ask two questions about the competitive objectives of R&D abroad: Will an organization be better off generating flows of product-process technology to the United States in order to compete effectively? Or will an organization be better off developing more suitable technology for growing foreign markets by promoting R&D abroad expressly for foreign markets? If the first question receives a strong positive response, the general manager must determine how realistic this foreign R&D

strategy is for the parent organization. Will the parent be able to finance the large expenditures necessary to create global product units? Is standardization of basic models or products feasible on an international scale? Can existing indigenous technology units be redirected and expanded to handle the projects? Does the company have large transfer technology units whose purpose can be redirected?

Many organizations will not be able to answer these questions satisfactorily. They may be better off conserving and consolidating R&D resources in the United States if new product-process technology must be developed for domestic commercialization. Other organizations may wish to pursue a strategy that fosters new and improved product-process work expressly for foreign markets. The R&D trade-offs of this strategy seem to revolve around the willingness of the parent to allow product-market diversifications abroad. Some organizations may wish to enter new, different businesses abroad. But if an enterprise is bent on following a concentrated global product-market strategy, it should be sure its goals are shared by foreign-based general managers with local R&D resources.

Of course, some organizations may wish to discover new opportunities and develop them into businesses that are not the same as domestic businesses. For these enterprises, the process of R&D abroad offers an alternative course of development. Indigenous technology units can sometimes make foreign subsidiaries technologically self-sufficient, while at the same time providing important product-process transfers of technology to other foreign nations.

At various times, corporate executives may want to ask another question that pertains specifically to the performance of exploratory research. Will the organizational system be better off performing long-term R&D activities abroad, and, if so, how should foreign corporate technology units be created and organized?

The evidence provided by this study suggests that corporate officials should be cautious about creating corporate technology units unless they are prepared to make a sizable, long-term investment; and the corporate technology unit operationally shares corporate R&D projects in the United States and/or with other foreign R&D units. A large investment is needed because the probability of successful projects (that is, those eventually resulting in development activity) appears to be quite low. The number of successes over a given time period (that is, one year) is related to the number of projects performed by R&D professionals. The higher the number of exploratory research projects, the higher the number of successes (but not the rate of success). R&D directors refer to this necessary condition as the high "critical mass factor" associated with exploratory research.

The evidence in this study suggests that the successful use of R&D resources may also depend upon focusing these resources within a limited number of related technological areas, and that this can be best achieved by limiting exploratory research to projects that support existing businesses, as opposed to exploratory research to develop new high-risk businesses. However, the ability to achieve a long-term focus of critical resources depends on maintaining continuity of R&D personnel. The need for a critical focus and continuity (especially within corporate technology units, where projects tend to be longer term) suggests that the national relocation of R&D personnel may be a short-sighted and unrealistic approach to staffing foreign R&D operations.

After a year or two, key R&D personnel may wish to return home. Also, the presence of a population of heterogeneous nationalities seems to provide an atmosphere of potential conflict, arising from different compensation expectations and other national differences. This kind of atmosphere may result in high turnover of personnel and a breakdown in the continuity of R&D operations.

Finally, the evidence suggests that R&D planners should integrate the planning of transfer technology units, indigenous technology units, and/or global product units with the planning of corporate technology units. The uncertainty and riskiness of exploratory research work suggest that wise contingency planning should include a fall-back position for a corporate technology unit, that is, the possibility of moving into another type of R&D activity. In addition, other foreign R&D units may need support, in the form of exploratory research, as their activities become more sophisticated. For both reasons, corporate decision makers should be wary about isolating corporate technology units from manufacturing and marketing units located abroad.

ROBERT RONSTADT is assistant professor of management at Babson College in Wellesley, Massachusetts, where he teaches courses in production and operations management, business policy, and entrepreneurship. He is also president of Robert Ronstadt Associates, Inc., a management research and consulting company.

Dr. Ronstadt was educated at the University of California at Berkeley (B.A. in history), the University of Oregon (M.S. in international studies), and at Harvard University (D.B.A.).

*THE MULTINATIONAL CORPORATION AND
SOCIAL CHANGE
edited by David E. Apter
Louis Wolf Goodman

FINANCIAL POLICIES FOR THE MULTINATIONAL
COMPANY: The Management of Foreign Exchange
Raj Aggarwal

MULTINATIONAL PRODUCT STRATEGY
A Typology for Analysis of Worldwide Product
Innovation and Diffusion
Georges Leroy

MARKETING MANAGEMENT IN MULTINATIONAL FIRMS
The Consumer Packaged Goods Industry
Ulrich E. Wiechmann

*Also available as a Praeger Special Studies Student Edition.